MEMOIRS
of the
American Mathematical Society

Number 448

The Metric Induced by the Robin Function

Norman Levenberg
Hiroshi Yamaguchi

July 1991 • Volume 92 • Number 448 (first of 4 numbers) • ISSN 0065-9266

American Mathematical Society
Providence, Rhode Island

1980 *Mathematics Subject Classification* (1985 *Revision*).
Primary 32F05; Secondary 31C10, 32F15.

Library of Congress Cataloging-in-Publication Data

Levenberg, Norman.
The metric induced by the Robin function/Norman Levenberg, Hiroshi Yamaguchi.
p. cm. - (Memoirs of the American Mathematical Society, ISSN 0065-9266; no. 448)
"July 1991, volume 92, number 448 (first of 4 numbers)."
Includes bibliographical references.
ISBN 0-8218-2520-8
1. Plurisubharmonic functions. 2. Pseudoconvex domains. I. Yamaguchi, Hiroshi, 1942- . II. Title. III. Title: Robin function. IV. Series.
QA3.A57 no. 448
[QA331.7]
510 s–dc20 91-13768
[515] CIP

Subscriptions and orders for publications of the American Mathematical Society should be addressed to American Mathematical Society, Box 1571, Annex Station, Providence, RI 02901-1571. *All orders must be accompanied by payment.* Other correspondence should be addressed to Box 6248, Providence, RI 02940-6248.

SUBSCRIPTION INFORMATION. The 1991 subscription begins with Number 438 and consists of six mailings, each containing one or more numbers. Subscription prices for 1991 are $270 list, $216 institutional member. A late charge of 10% of the subscription price will be imposed on orders received from nonmembers after January 1 of the subscription year. Subscribers outside the United States and India must pay a postage surcharge of $25; subscribers in India must pay a postage surcharge of $43. Expedited delivery to destinations in North America $30; elsewhere $82. Each number may be ordered separately; *please specify number* when ordering an individual number. For prices and titles of recently released numbers, see the New Publications sections of the NOTICES of the American Mathematical Society.

BACK NUMBER INFORMATION. For back issues see the AMS Catalogue of Publications.

MEMOIRS of the American Mathematical Society (ISSN 0065-9266) is published bimonthly (each volume consisting usually of more than one number) by the American Mathematical Society at 201 Charles Street, Providence, Rhode Island 02904-2213. Second Class postage paid at Providence, Rhode Island 02940-6248. Postmaster: Send address changes to Memoirs of the American Mathematical Society, American Mathematical Society, Box 6248, Providence, RI 02940-6248.

10 9 8 7 6 5 4 3 2 1 95 94 93 92 91

TABLE OF CONTENTS

ABSTRACT

Let $D \subset\subset \mathbb{C}^n$ be a domain with C^2 boundary. For each point $\xi \in D$, we have the Green function $G(\xi, z)$ associated to the Laplace operator Δ with pole at ξ and the Robin constant $\Lambda(\xi) \equiv \lim_{z \to \xi} \left[G(\xi, z) - \frac{1}{\|z-\xi\|^{2n-2}} \right]$ for (D, ξ). As ξ varies in D, we get a function, the Robin function $\Lambda = \Lambda(\xi)$ for D. We show that if D is pseudoconvex, then the functions $-\Lambda(\xi)$ and $\log(-\Lambda(\xi))$ are real analytic, strictly plurisubharmonic exhaustion functions for D. Moreover, for an arbitrary smoothly bounded domain D given by a defining function ψ, i.e., $D = \{\xi \in \mathbb{C}^n : \psi(\xi) < 0\}$, we prove the following boundary regularity of Λ : the function $\Lambda(\xi)\psi(\xi)^{2n-2}$ is C^2 up to ∂D.

We then study the Kähler metric

$$ds^2 = \sum_{\alpha, \beta=1}^{n} \frac{\partial^2 \log(-\Lambda)}{\partial \xi_\alpha \, \partial \bar{\xi}_\beta} (\xi) \, d\xi_\alpha \otimes d\bar{\xi}_\beta$$

associated to a pseudoconvex domain D with C^2 boundary. Using computations involving the complex Hessian of $\log(-\Lambda(\xi))$ and the above boundary asymptotics of $\Lambda(\xi)$, we show that for a wide class of domains, including the strictly pseudoconvex and R^{2n}-convex ones, this metric is complete. A brief discussion of the relationship between the Laplacian of $-\Lambda(\xi)$ and the Bergman kernel function for D is also given.

0. INTRODUCTION

Given a domain $D \subset R^m$ $(m > 2)$ with smooth boundary and given a point $\xi \in D$, the Green function $G(\xi,z)$ for D with pole at ξ is the unique function of $z \in D$ which is harmonic in $D - \{\xi\}$, tends to 0 at ∂D, and satisfies

$$G(\xi,z) - \frac{1}{\|z-\xi\|^{m-2}}$$

is harmonic for z near ξ. The number

$$\Lambda(\xi) \equiv \lim_{z\to\xi} \left[G(\xi,z) - \frac{1}{\|z-\xi\|^{m-2}} \right]$$

is called the Robin constant for (D,ξ). Letting the point ξ vary in D, we get a negative valued, real analytic function defined on D, the Robin function $\Lambda = \Lambda(\xi)$. This function, which tends to $-\infty$ near ∂D, arises in studying the classical (Newtonian) potential theory associated with the Laplace operator Δ on D.

If $m = 2n$ and we identify R^{2n} with $\mathbb{C}^n$, a surprising result of the second author relates the R^{2n}-potential theory with pluripotential theory (i.e., the study of plurisubharmonic functions) in $\mathbb{C}^n$. His result is that if D is pseudoconvex, then the functions $-\Lambda(\xi)$ and $\log(-\Lambda(\xi))$ are strictly plurisubharmonic in D. The main idea used in [Y] to prove this result was to use the technique of variation of domains in $\mathbb{C}^n$: if $D(t)$ is a domain in $\mathbb{C}^n$ for each t in a disc $B = \{t \in \mathbb{C}: |t| < \rho\}$, one studies how various function-theoretic quantities associated to the domains $D(t)$

vary with the parameter $t \in B$. In this particular case, if $\xi \in D(t)$ for all $t \in B$, we consider the Robin constant $\lambda(t)$ for $(D(t),\xi)$ and consider $\lambda = \lambda(t)$ as a function on B. If the set

$$\mathcal{D} \equiv \{(t,z) \in \mathbb{C} \times \mathbb{C}^n : t \in B, \quad z \in D(t)\}$$

is a pseudoconvex domain in $B \times \mathbb{C}^n$, then $-\lambda(t)$ and $\log(-\lambda(t))$ are subharmonic functions in B.

The technique of variation of domains was used by Oka in his study of pseudoconvex domains in $\mathbb{C}^2$ [OK] and later by Nishino in his work on value distribution theory of entire functions in two variables [N]. These latter works, together with Rauch's paper [R] on variation of moduli for compact Riemann surfaces, served as inspiration for the second author in [Y_1] where he showed that for a variation of Riemann surfaces D(t) over $\mathbb{C}$, if the total space $\mathcal{D} \equiv \bigcup_{t \in B} (t, D(t))$ over $B \times \mathbb{C}$ is a two-dimensional Stein manifold, then $-\lambda(t)$ is subharmonic in B where $\lambda(t)$ is the Robin constant for $(D(t),\xi)$ associated to the Green function for $D(t)$ with logarithmic singularity at ξ (if the Green function for $(D(t),\xi)$ does not exist, i.e., $D(t)$ is parabolic, then we set $\lambda(t) = +\infty$). This result was used to study the level sets $\sum_c \equiv \{(z_1,z_2) \in \mathbb{C}^2 : f(z_1,z_2) = c\}$ of a non-constant entire function f in $\mathbb{C}^2$. In particular, if $E \equiv \{c \in \mathbb{C}: \sum_c$ has at least one parabolic connected component$\}$ has positive logarithmic capacity, then every component of $\sum_c$ is parabolic for all $c \in \mathbb{C}$.

In this paper, we give a new proof of Yamaguchi's $\mathbb{C}^n$-result $(n > 1)$ concerning the Robin function $\Lambda(\xi)$ associated to a pseudoconvex domain D. We derive a formula for the Laplacian $\dfrac{\partial^2\lambda}{\partial t\, \partial \bar{t}}$ of the Robin function $\lambda(t)$

associated to a variation of domains $D(t) \subset \mathbb{C}^n$; this leads to an explicit expression for the complex Hessian of the functions $-\Lambda(\xi)$ and $\log(-\Lambda(\xi))$ where $\Lambda(\xi)$ is the Robin function for (D,ξ). The proof is entirely self-contained, and, we hope, sheds some light on the curious relationship between the R^{2n} - and $\mathbb{C}^n$-potential theories. In addition, we study the boundary asymptotics of the Robin function $\Lambda(\xi)$ for arbitrary smoothly bounded domains and we later use these results, together with the explicit formula for the complex Hessian of $\log(-\Lambda(\xi))$, to study a natural metric induced by this function, namely $ds^2 \equiv \sum_{\alpha,\beta=1}^{n} \frac{\partial^2 \log(-\Lambda)}{\partial\xi_\alpha \partial\bar{\xi}_\beta}(\xi)\, d\xi_\alpha \otimes d\bar{\xi}_\beta$, $\xi \in D$. We call ds^2 the Λ-metric for D. For a large class of domains, including the strictly pseudoconvex and R^{2n}-convex ones, this metric is shown to be complete. It is unknown if ds^2 is complete in D for an arbitrary bounded pseudoconvex domain with smooth boundary. Many other open questions involving the Robin function $\Lambda(\xi)$ and the Λ-metric ds^2 are discussed throughout the paper.

The outline of presentation is as follows.

In Chapter 1, we define the Levi-curvature $k_2(t,z)$ associated to certain boundary points of domains $D \subset \mathbb{C} \times \mathbb{C}^n$. This quantity will occur in our fundamental formula for the Laplacian $\frac{\partial^2\lambda}{\partial t\partial\bar{t}}$ of the Robin function $\lambda(t)$ associated to a smooth variation of domains in $\mathbb{C}^n$ in Chapter 2. In Chapters 3-6, we make a detailed study of the boundary behavior of the Robin function $\Lambda(\xi)$ associated to a domain $D \subset \mathbb{C}^n$ with smooth boundary. Chapters 4-5 are rather technical and may be omitted upon first reading. In Chapter 7, we use the fundamental formula from Chapter 2 and a differential inequality characterizing pseudoconvex boundary points to give a new proof of Theorem 8.1 [Y] that $-\Lambda(\xi)$ and $\log(-\Lambda(\xi))$ are strictly plurisubharmonic when D is a bounded pseudoconvex domain with smooth boundary. We then briefly discuss, in Chapter 8, the relationship between the Laplacian of $-\Lambda(\xi)$ for a bounded, smoothly bounded pseudoconvex domain D and the Bergman kernel function $K(\xi,z)$ for D restricted to the diagonal $z = \xi$. Finally, for such domains, we study properties of the Kähler metric $ds^2 = \sum_{\alpha,\beta=1}^{n} \frac{\partial^2 \log(-\Lambda)}{\partial\xi_\alpha \partial\bar{\xi}_\beta}(\xi) d\xi_\alpha \otimes d\bar{\xi}_\beta$ in the remaining five chapters. Included are some explicit calculations for some specific examples.

A remark on notation: $\|v\|$ will denote the Euclidean norm of a vector v in $R^{2n} = \mathbb{C}^n$ if $n > 1$; for $n = 1$, we write $|v|$. Also, if $\psi = \psi(z_1,\ldots,z_n)$ is a real-valued function of $z = (z_1,\ldots,z_n) \in \mathbb{C}^n$, we write $\text{Grad}\ \psi = \text{Grad}_{(z)}\psi = \left(\frac{\partial\psi}{\partial z_1},\ldots,\frac{\partial\psi}{\partial z_n}\right)$. Thus we identify $2\ \overline{\text{Grad}\ \psi}$ with the usual R^{2n}-gradient $\left(\frac{\partial\psi}{\partial x_1}, \frac{\partial\psi}{\partial y_1},\ldots,\frac{\partial\psi}{\partial x_n}, \frac{\partial\psi}{\partial y_n}\right)$ where $z_j = x_j + iy_j$, $j = 1,\ldots,n$.

Acknowledgements. The authors are greatly indebted to a long list of friends and colleagues. To avoid the danger of omitting a name from the list, we offer instead a general thank-you to everyone who has contributed to our work, and a special thank-you to the referee for valuable comments.

1. LEVI-CURVATURE

Let D be a domain in $\mathbb{C}^2$. If ∂D is of class C^2, we can find a real-valued C^2 defining function ψ for D, i.e., ψ is of class C^2 in a neighborhood of D and $D = \{(t,z) \in \mathbb{C}^2 : \psi(t,z) < 0\}$, $\partial D = \{(t,z) \in \mathbb{C}^2 : \psi(t,z) = 0\}$, with $\mathrm{Grad}_{(t,z)}\psi = \left(\frac{\partial\psi}{\partial t}, \frac{\partial\psi}{\partial z}\right) \neq (0,0)$ on ∂D. Thus for each $(t_0, z_0) \in \partial D$, the vector $\left(\frac{-\partial\psi}{\partial z}(t_0,z_0), \frac{\partial\psi}{\partial t}(t_0,z_0)\right)$ spans the complex tangent space to ∂D at (t_0,z_0). Hence D is pseudoconvex at $\frac{\partial\psi}{\partial z}(t_0,z_0)$ precisely when the Levi-form

$$(1.1) \qquad L_{(t,z)}\psi \equiv \frac{\partial^2\psi}{\partial t\partial\bar{t}}\left|\frac{\partial\psi}{\partial z}\right|^2 - 2\,\mathrm{Re}\left\{\frac{\partial\psi}{\partial t}\,\frac{\partial\psi}{\partial\bar{z}}\,\frac{\partial^2\psi}{\partial\bar{t}\partial z}\right\} + \frac{\partial^2\psi}{\partial z\partial\bar{z}}\left|\frac{\partial\psi}{\partial t}\right|^2$$

is nonnegative at (t_0,z_0). In general, for a C^2 function ψ defined in a neighborhood of D, we will call the quantity $L_{(t,z)}\psi$ in (1.1) the <u>Levi-form of ψ with respect to (t,z)</u>.

More generally, if D is a domain in $\mathbb{C}^{n+1} = \mathbb{C} \times \mathbb{C}^n$ and $\psi = \psi(t,z) \equiv \psi(t, z_1, \cdots, z_n)$ is a real-valued function of class C^2 in a neighborhood of D, we call

$$(1.2) \qquad \mathcal{L}\psi = \sum_{i=1}^{n} L_{(t,z_i)}\psi = \frac{\partial^2\psi}{\partial t\partial\bar{t}}\|\mathrm{Grad}_{(z)}\psi\|^2 - 2\,\mathrm{Re}\left\{\frac{\partial\psi}{\partial t}\sum_{i=1}^{n}\frac{\partial\psi}{\partial\bar{z}_i}\,\frac{\partial^2\psi}{\partial\bar{t}\partial z_i}\right\} + \left|\frac{\partial\psi}{\partial t}\right|^2\Delta_{(z)}\psi$$

where $\mathrm{Grad}_{(z)}\psi = \left(\frac{\partial\psi}{\partial z_1}, \cdots, \frac{\partial\psi}{\partial z_n}\right)$, $\Delta_{(z)}\psi = \sum_{i=1}^{n}\frac{\partial^2\psi}{\partial z_i\partial\bar{z}_i}$, and $\|\cdot\|$ denotes the Euclidean norm of $\mathbb{C}^n = \mathbb{R}^{2n}$, the <u>Levi-form of ψ with respect to t and $z = (z_1, \cdots, z_n)$</u>. Note that if ψ is a C^2 defining function for D in the sense described above, $\mathcal{L}\psi$ is <u>not</u> the usual Levi-form applied to a complex tangent vector; rather, it is the <u>sum</u> of the usual Levi-form applied to the n complex tangent vectors $\left(\frac{-\partial\psi}{\partial z_i}, 0, \cdots, \frac{\partial\psi}{\partial t}, 0, \cdots, 0\right)$.

Received by the editor April 20, 1989 and in revised form June 15, 1990.

To define the Levi-curvature $k_2(t,z)$ described in the introduction, we consider domains $D \subset \mathbb{C} \times \mathbb{C}^n$ given by a defining function ψ satisfying an additional property. Precisely, we assume that $D \subset B \times \Omega$ where $B \subset C$, $\Omega \subset \mathbb{C}^n$ are domains and that not only is $\psi(t,z)$ of class C^2 for $t \in B$ — not necessarily for $t \in \partial B$ — but also, for each fixed $t \in B$, $\text{Grad}_{(z)}\psi = \left(\frac{\partial\psi}{\partial z_1}, \cdots, \frac{\partial\psi}{\partial z_n}\right) \neq (0,\cdots,0)$ for $z \in \partial D(t) = \{z \in \Omega : \psi(t,z) = 0\}$. (We remark that we are looking ahead to Chapter 2 where we study the variation of domains $D(t) \equiv \{z \in \Omega : \psi(t,z) < 0\}$ for $t \in B$). If $\varphi(t,z)$ is another such defining function for D on $B \times \Omega$, then $\varphi = f\psi$ where f is a C^2 function which is positive on $\bar{S} \equiv \{(t,z) : t \in B,\ z \in \partial D(t)\}$. Thus $\text{Grad}_{(t,z)}\varphi = f \cdot \text{Grad}_{(t,z)}\psi$ on S; in particular, $\frac{\partial\varphi}{\partial t} = f\,\frac{\partial\psi}{\partial t}$ and $\|\text{Grad}_{(z)}\varphi\| = |f|\ \|\text{Grad}_{(z)}\psi\|$ so that, by direct calculation, $\mathcal{L}\varphi = f^3\mathcal{L}\psi$ on S and the quantities

$$k_1(t,z) \equiv \frac{\partial\psi}{\partial t}\Big/\|\text{Grad}_{(z)}\psi\|$$

(1.3) and

$$k_2(t,z) \equiv \mathcal{L}\psi\Big/\|\text{Grad}_{(z)}\psi\|^3$$

are well-defined functions on S which are independent of the choice of defining function for D. We call $k_2(t,z)$ the <u>Levi-curvature of ∂D at (t,z)</u>.

We have the following simple but important property of the Levi-curvature : it is invariant in the z-variables under unitary transformations of $\mathbb{C}^n$. More generally, we have the following result.

<u>Proposition 1.1.</u> Let $D = \{(t,z) \in \mathbb{C} : \psi(t,z) < 0\}$ be a domain as above. Let $T : \mathbb{C}^{n+1} \to \mathbb{C}^{n+1}$ be a transformation of the form

$$T(t,z) = (t,w) = (t,\ cAz + b)$$

where $c \in \mathbb{C} - \{0\}$, $b \in \mathbb{C}^n$, and A is an $n \times n$ unitary transformation. Then the Levi-curvature $k_2^*(t,w)$ for $T(D)$ satisfies $k_2^*(t,w) = |c|\ k_2(t,z)$ for $(t,w) \in T(S)$.

Proof. Note that $T(D) = \{(t,w) : \psi^*(t,w) < 0\}$ where $\psi^*(t,w) = \psi(t,z)$. Clearly $\|\mathrm{Grad}_{(z)}\psi(t,z)\| = |c|\ \|\mathrm{Grad}_{(w)}\psi^*(t,w)\|$ and $\mathcal{L}\psi(t,z) = |c|^2\ \mathcal{L}\psi^*(t,w)$; hence $k_2^*(t,w) = |c|\ k_2(t,z)$. Setting $b = 0$ and $c = 1$ yields the invariance of $k_2(t,z)$ in the z variables. ■

Note that the sign of $k_2(t,z)$ is invariant for the transformations described in Proposition 1.1. There is an obvious sufficient condition on D guaranteeing nonnegativity of the Levi-curvature: pseudoconvexity.

Proposition 1.2. Let D be as above.

(a) If D is pseudoconvex at $(t_0, z_0) \in S$, then $k_2(t_0, z_0) \geq 0$.

(b) If D is strictly pseudoconvex at $(t_0, z_0) \in S$, then $k_2(t_0, z_0) > 0$.

(c) There exist domains $D \subset B \times \Omega$ satisfying $k_2(t,z) > 0$ for all $(t,z) \in S$ which are not pseudoconvex.

Remark. Part (c) shows, as cautioned earlier, that there is a difference between the Levi-form $\mathcal{L}\psi$ and the usual Levi-form when $n > 1$.

Proof. (a) If D is pseudoconvex at (t_0, z_0), then $L_{(t,z_i)}\psi(t_0, z_0) \geq 0$, $i = 1, \cdots, n$, so that $\mathcal{L}\psi(t_0, z_0) \geq 0$ and hence $k_2(t_0, z_0) \geq 0$..

(b) If D is strictly pseudoconvex at (t_0, z_0), then, since $\mathrm{Grad}_{(z)}\psi(t_0, z_0) \neq (0, \cdots, 0)$, there exists $i \in \{1, \cdots, n\}$ such that $L_{(t,z_i)}\psi(t_0, z_0) > 0$. Hence $\mathcal{L}\psi(t_0, z_0) \geq L_{(t,z_i)}\psi(t_0, z_0) > 0$ so that $k_2(t_0, z_0) > 0$.

(c) We first mention that to get non-pseudoconvex domains satisfying $k_2(t,z) \geq 0$ for all $(t,z) \in S$ is easy. For if we start with any domain $\Omega \subset \mathbb{C}^n$, then a defining function ψ for the product domain $D \equiv B \times \Omega$ satisfies $\frac{\partial\psi}{\partial t} \equiv 0$ on S which gives $\mathcal{L}\psi \equiv 0$ and $k_2(t,z) \equiv 0$ on S even if Ω, and hence D, is not pseudoconvex. To give an example of a non-pseudoconvex $D \subset B \times \Omega$ with $k_2(t,z) > 0$ on S, take

$$D \equiv \{(t,z_1,z_2) \in \mathbb{C}^3 : |t| < \tfrac{1}{20}, \psi(t,z_1,z_2) \equiv \tfrac{1}{3} + |z_1|^2 - \frac{1}{1+|z_2|^2} + 2\,\mathrm{Re}(t) < 0\}.$$

By explicit calculation, $L_{(t,z_1)}\psi \equiv 1$ and $L_{(t,z_2)}\psi \equiv \dfrac{1-|z_2|^2}{(1+|z_2|^2)^3}$ so that $\mathcal{L}\psi \geq \dfrac{1+2|z_2|^2}{(1+|z_2|^2)^3} > 0$ (note that $\frac{\partial\psi}{\partial t} \equiv 1$ and $\mathrm{Grad}_{(z)}\psi = \left(\bar{z}_1, \dfrac{\bar{z}_2}{(1+|z_2|^2)^2}\right) \neq (0,0)$ on $\partial D(t)$ (since $|t| < \frac{1}{20}$) so that ψ is a valid defining function for D as described above). Finally, for each t, if $z_1 = 0$ then $(0,z_2) \in \partial D(t)$ implies that $|z_2|^2 = \dfrac{1}{2\,\mathrm{Re}(t) + \frac{1}{3}} - 1 \equiv F(t)$, and, for each t, $\frac{17}{13} < F(t) < \frac{23}{7}$. In particular, each $\partial D(t)$ contains a point $(0,z_2)$ with $|z_2| > 1$ and $L_{(t,z_2)}\psi < 0$ at such points. Hence $\{(t,z_1,z_2) \in D : z_1 = 0\}$ is not pseudoconvex in $\mathbb{C}^2$; thus D is not pseudoconvex. Note that the domain D in this example is bounded. ∎

2. SMOOTH VARIATION OF DOMAINS

In this section, we analyze the inequality in Lemma 3.1 [Y] to obtain an explicit formula for the Laplacian $\frac{\partial^2\lambda}{\partial t\partial\bar{t}}$ of the Robin function associated to a variation of smooth domains in $\mathbb{C}^n$ with a constant section. We recall the general set-up from [Y], Section 3. Let $\mathcal{D}$ be a domain in $B \times \mathbb{C}^n$ where B is a domain in $\mathbb{C}$ (in all that follows, we could consider the more general situation where $\mathcal{D}$ is an unramified domain over $\mathbb{C}^n$; in this paper, for simplicity, we prove our results for $\mathcal{D} \subset B \times \mathbb{C}^n$). For each $t \in B$, we call $D(t) \equiv \{z \in \mathbb{C}^n : (t,z) \in \mathcal{D}\}$ the <u>fiber of $\mathcal{D}$ over t</u> and we consider $\mathcal{D} = \bigcup_{t\in B} (t, D(t))$ as a variation of open sets $D(t) \subset \mathbb{C}^n$ with parameter space $B \subset \mathbb{C}$; we often write $\mathcal{D} : t \to D(t)$. Later we will have occasion to use variations with parameter space $B \subset \mathbb{C}^n$; the modifications in this case in the definition of a smooth variation, to be defined below, are clear. In this section, we assume $\mathcal{D} : t \to D(t)$ satisfies the following three conditions.

<u>Condition 2.1.</u> There exists a domain $\tilde{\mathcal{D}} \supset \mathcal{D}$ with $\tilde{\mathcal{D}} \subset B \times \mathbb{C}^n$ and a C^∞ function $\psi(t,z)$ in $\tilde{\mathcal{D}}$ which defines $\mathcal{D}$, i.e.,

$$\mathcal{D} = \{(t,z) \in \tilde{\mathcal{D}} : \psi(t,z) < 0\}$$

$$\mathcal{S} \equiv \{(t,z) : t \in B,\ z \in \partial D(t)\} = \{(t,z) \in \tilde{\mathcal{D}} : \psi(t,z) = 0\}$$

and $\mathrm{Grad}_{(t,z)}\psi \neq (0,\cdots,0)$ on $\mathcal{S}$.

<u>Condition 2.2.</u> For each $t \in B$, $D(t) \subset\subset \tilde{D}(t)$ and $\psi(t,z)$ is a defining function for $D(t)$, i.e.,

$$D(t) = \{z \in \tilde{D}(t) : \psi(t,z) < 0\}$$

$$\partial D(t) = \{z \in \tilde{D}(t) : \psi(t,z) = 0\}$$

and $\mathrm{Grad}_{(z)}\psi \neq (0,\cdots,0)$ on $\partial D(t)$.

We call $\mathcal{D} : t \to D(t)$ a smooth variation of domains in $\mathbb{C}^n$ and say that the double $(\tilde{\mathcal{D}}, \psi)$ defines $\mathcal{D}$.

Condition 2.3. There exists a point $\xi_0 \in \mathbb{C}^n$ such that $B \times \{\xi_0\} \subset \mathcal{D}$, i.e., the variation $\mathcal{D} : t \to D(t)$ has a constant section $\xi(t) \equiv \xi_0$.

In this situation, we can construct, for each domain $D(t)$, the Green function $g(t,z)$ with pole at ξ_0:

(i) $g(t,z)$ is harmonic for $z \in D(t) - \{\xi_0\}$

(ii) $g(t,z)$ is continuous up to $\partial D(t)$ and $g(t,z) = 0$ for $z \in \partial D(t)$

(iii) in a neighborhood of ξ_0, we have

$$g(t,z) = \frac{1}{\|z - \xi_0\|^{2n-2}} + \lambda(t) + h(t,z) \tag{2.1}$$

where $\lambda(t)$ is the Robin constant for $(D(t), \xi_0)$, $h(t,z)$ is harmonic for $z \in D(t)$ and

$$h(t,\xi_0) = 0. \tag{2.2}$$

For each $t \in B$, we thus get a Robin constant $\lambda(t)$; we call $\lambda(t)$ the Robin function associated to $\mathcal{D}$.

Remark. The above Robin function arises from a variation of domains in $\mathbb{C}^n$ with fixed pole; later we will study a Robin function associated to a fixed domain in $\mathbb{C}^n$ which arises by varying the pole. Neither of these Robin functions is related to the one studied by Bedford and Taylor [BT].

For a smooth variation $\mathcal{D}$, $\mathcal{S} = \bigcup_{t \in B} (t, \partial D(t))$ and the variation $\mathcal{D} \cup \mathcal{S} : t \mapsto D(t) \cup \partial D(t) = \overline{D(t)}$ is locally diffeomorphically equivalent to the trivial variation $B \times (D(t_0) \cup \partial D(t_0))$; i.e., for each $t_0 \in B$, there exists a neighborhood $B_0 \subset B$ of t_0 such that $\mathcal{D} \cup \mathcal{S} \big|_{B_0}$

$\left(= \bigcup_{t \in B_0} (t, \overline{D(t)}) \right)$ is locally diffeomorphically equivalent to $B_0 \times \overline{D(t_0)}$.

It follows that $g(t,z)$ has a C^4 extension to a neighborhood of $\mathcal{D} - B \times \{\xi_0\}$ and thus $\lambda(t)$ is a C^4 function on B (see Proposition 3.1. [Y]). Furthermore, since ξ_0 is constant, for each $t \in B$, $\frac{\partial g}{\partial t}(t,z)$ and $\frac{\partial^2 g}{\partial t \partial \bar{t}}(t,z)$ are harmonic in all of $D(t)$, even at ξ_0, and

$$(2.3) \qquad \frac{\partial g}{\partial t}(t,\xi_0) = \frac{\partial \lambda}{\partial t}(t) \ ; \quad \frac{\partial^2 g}{\partial t \partial \bar{t}}(t,\xi_0) = \frac{\partial^2 \lambda}{\partial t \partial \bar{t}}(t) \ , \qquad t \in B.$$

Given $\xi \in D(t)$, we let $g_\xi(t,z)$ denote the Green function for $(D(t),\xi)$ (note that $g_{\xi_0}(t,z) = g(t,z)$). The main theorem of this section is the following.

<u>Theorem 2.1.</u> For all $\xi \in D(t)$, $t \in B$,

$$(2.4) \quad \frac{\partial g}{\partial t}(t,\xi) = \frac{1}{2(n-1)w_{2n}} \int_{\partial D(t)} k_1(t,z) \, \|\mathrm{Grad}_{(z)} g(t,z)\| \, \frac{\partial g_\xi(t,z)}{\partial n_z} \, dS_z$$

$$(2.5) \quad \frac{\partial^2 g}{\partial t \partial \bar{t}}(t,\xi) = \frac{1}{2(n-1)w_{2n}} \int_{\partial D(t)} k_2(t,z) \, \|\mathrm{Grad}_{(z)} g(t,z)\| \, \frac{\partial g_\xi(t,z)}{\partial n_z} \, dS_z$$

$$- \frac{4}{(n-1)w_{2n}} \operatorname{Re} \iint_{D(t)} \left[\sum_{\alpha=1}^{n} \frac{\partial^2 g(t,z)}{\partial t \partial \bar{z}_\alpha} \, \frac{\partial^2 g_\xi(t,z)}{\partial \bar{t} \partial z_\alpha} \right] dV_z$$

where w_{2n} = surface area of the unit sphere in $\mathbb{C}^n = \mathbb{R}^{2n}$, dS_z = surface area on $\partial D(t)$, dV_z = Lebesgue measure in $\mathbb{C}^n = \mathbb{R}^{2n}$, and $\frac{\partial}{\partial n_z}$ = outer normal derivative on $\partial D(t)$.

Proof. Fix $t \in B$. Since $\frac{\partial g}{\partial t}(t,z)$ and $\frac{\partial^2 g}{\partial t \partial \bar{t}}(t,z)$ are harmonic for $z \in D(t)$ and continuous up to $\partial D(t)$, we have

$$\frac{\partial g}{\partial t}(t,\xi) = \frac{-1}{2(n-1)w_{2n}} \int_{\partial D(t)} \frac{\partial g}{\partial t}(t,z) \frac{\partial g_\xi}{\partial n_z}(t,z)dS_z \tag{2.6}$$

$$\frac{\partial^2 g}{\partial t \partial \bar{t}}(t,\xi) = \frac{-1}{2(n-1)w_{2n}} \int_{\partial D(t)} \frac{\partial^2 g}{\partial t \partial \bar{t}}(t,z) \frac{\partial g_\xi}{\partial n_z}(t,z)dS_z \tag{2.7}$$

for all $\xi \in D(t)$. Since $g(t,z) > 0$ on $\mathcal{D}$, $g(t,z) = 0$ on $\mathcal{S}$ and $\|\mathrm{Grad}_{(z)}\, g(t,z)\| = -\frac{1}{2}\frac{\partial g}{\partial n_z}(t,z) > 0$ on $\mathcal{S}$, we can use $-g(t,z)$ as a defining function for $\mathcal{D}$ and by (1.3) we have

$$\frac{\partial g}{\partial t}(t,z) = -k_1(t,z)\ \|\mathrm{Grad}_{(z)} g(t,z)\| \tag{2.8}$$

and

$$\mathcal{L}g(t,z) = -k_2(t,z)\ \|\mathrm{Grad}_{(z)} g(t,z)\|^3 \tag{2.9}$$

for all $(t,z) \in \mathcal{S}$. Inserting (2.8) into (2.6) immediately yields (2.4). To prove (2.5), note that since $g(t,z)$ is harmonic for $z \in D(t) - \{\xi_0\}$ and is of class C^4 up to $\partial D(t)$, $\Delta_{(z)} g(t,z) = 0$ for $z \in \partial D(t)$. Thus by (1.2) and (2.9),

$$\frac{\partial^2 g}{\partial t \partial \bar{t}} = -k_2 \|\mathrm{Grad}_{(z)} g\| + 2\mathrm{Re}\left\{ \sum_{\alpha=1}^{n} \frac{\frac{\partial g}{\partial t}}{\|\mathrm{Grad}_{(z)} g\|} \frac{\partial^2 g}{\partial z_\alpha \partial \bar{t}} \frac{\overline{\frac{\partial g}{\partial \bar{z}_\alpha}}}{\|\mathrm{Grad}_{(z)} g\|} \right\} \tag{2.10}$$

for $z \in \partial D(t)$. But $g_\xi(t,z) = g(t,z) = 0$ for $z \in \partial D(t)$ and

$$\|\mathrm{Grad}_{(z)} g_\xi(t,z)\| = -\frac{1}{2}\frac{\partial g_\xi}{\partial n_z}(t,z)\ ,\quad z \in \partial D(t)\ , \tag{2.11}$$

thus replacing $\dfrac{\frac{\partial g}{\partial t}}{\|\mathrm{Grad}_{(z)} g\|}$ by $\dfrac{\frac{\partial g_\xi}{\partial t}}{\|\mathrm{Grad}_{(z)} g_\xi\|}$ in (2.10) and then putting (2.10) into (2.7) we obtain

$$(2.12)\quad \frac{\partial^2 g}{\partial t\partial\bar{t}}(t,\xi) = \frac{1}{2(n-1)w_{2n}}\int_{\partial D(t)} k_2(t,z)\|\mathrm{Grad}_{(z)} g(t,z)\| \frac{\partial g_\xi}{\partial n_z}(t,z)dS_z$$

$$+\frac{2}{(n-1)w_{2n}}\mathrm{Re}\left\{\sum_{\alpha=1}^{n}\int_{\partial D(t)}\left[\frac{\partial g_\xi}{\partial t}(t,z)\ \frac{\partial^2 g}{\partial\bar{t}\partial z_\alpha}(t,z)\ \frac{\frac{\partial g}{\partial \bar{z}_\alpha}(t,z)}{\|\mathrm{Grad}_{(z)} g(t,z)\|}\right]dS_z\right\}$$

$$\equiv I + J .$$

Since

$$\left(\frac{\partial g}{\partial\bar{z}_\alpha}(t,z)/\|\mathrm{Grad}_{(z)} g(t,z)\|\right)dS_z = \frac{i^n}{2^{n-1}}dz_1\wedge d\bar{z}_1\wedge\ldots dz_\alpha\wedge\widehat{d\bar{z}_\alpha}\ldots\wedge dz_n\wedge d\bar{z}_n$$

for $z\in\partial D(t)$ (see Proposition 3.3 [Y]), we get

$$J = \frac{2}{(n-1)w_{2n}}\frac{1}{2^{n-1}}\mathrm{Re}\left\{i^n\sum_{\alpha=1}^{n}\int_{\partial D(t)}\frac{\partial g_\xi}{\partial t}\frac{\partial^2 g}{\partial\bar{t}\partial z_\alpha}\right.$$

$$\left. dz_1\wedge d\bar{z}_1\wedge\ldots dz_\alpha\wedge\widehat{d\bar{z}_\alpha}\ldots\wedge dz_n\wedge d\bar{z}_n\right\}$$

$$= \frac{-1}{2^{n-2}(n-1)w_{2n}}\mathrm{Re}\left\{i^n\sum_{\alpha=1}^{n}\iint_{D(t)}\frac{\partial}{\partial\bar{z}_\alpha}\left(\frac{\partial g_\xi}{\partial t}\frac{\partial^2 g}{\partial\bar{t}\partial z_\alpha}\right)dz_1\wedge d\bar{z}_1\wedge\ldots\wedge dz_n\wedge d\bar{z}_n\right\}$$

$$= \frac{-4}{(n-1)w_{2n}}\mathrm{Re}\left\{\sum_{\alpha=1}^{n}\iint_{D(t)}\left[\frac{\partial^2 g_\xi}{\partial t\partial\bar{z}_\alpha}\frac{\partial^2 g}{\partial\bar{t}\partial z_\alpha}+\frac{\partial g_\xi}{\partial t}\frac{\partial^3 g}{\partial\bar{t}\partial z_\alpha\partial\bar{z}_\alpha}\right]dV_z\right\}$$

where we have used Stokes theorem for the second equality. Since $\frac{\partial g}{\partial\bar{t}}(t,z)$

is harmonic for $z \in D(t)$, $\sum_{\alpha=1}^{n} \frac{\partial^3 g}{\partial\bar{t}\partial z_\alpha \partial\bar{z}_\alpha}(t,z) = 0$ for $z \in D(t)$ and

$$J = \frac{-4}{(n-1)w_{2n}} \operatorname{Re}\left\{ \sum_{\alpha=1}^{n} \iint_{D(t)} \left[\frac{\partial^2 g_\xi}{\partial t \partial \bar{z}_\alpha} \frac{\partial^2 g}{\partial \bar{t} \partial z_\alpha} dV_z \right\} .$$

Using this in (2.12) gives (2.5). ∎

Setting $\xi = \xi_0$ in (2.4) and (2.5) and using equations (2.3) and (2.11), we obtain the fundamental formula for $\frac{\partial\lambda}{\partial t}$ and for $\frac{\partial^2\lambda}{\partial t\partial\bar{t}}$.

Corollary 2.1. (Fundamental formula). For $t \in B$,

$$\text{(2.13)} \qquad \frac{\partial\lambda}{\partial t}(t) = \frac{-1}{(n-1)w_{2n}} \int_{\partial D(t)} k_1(t,z) \|\mathrm{Grad}_{(z)} g\|^2 dS_z$$

and

$$\text{(2.14)} \qquad \frac{\partial^2\lambda}{\partial t\partial\bar{t}}(t) = \frac{-1}{(n-1)w_{2n}} \int_{\partial D(t)} k_2(t,z) \|\mathrm{Grad}_{(z)} g\|^2 dS_z - \frac{4}{(n-1)w_{2n}} \iint_{D(t)} \sum_{\alpha=1}^{n} \left| \frac{\partial^2 g}{\partial t \partial \bar{z}_\alpha} \right|^2 dV_z .$$

We now see the significance of the sign of the Levi-curvature.

Corollary 2.2. If $k_2(t,z) \geq 0$ for all $(t,z) \in S$, then $\lambda(t)$ is superharmonic on B.

In particular, from Proposition 1.2, we have the following important result (cf. Lemma 3.1 [Y]).

Corollary 2.3. If $\mathcal{D}$ is pseudoconvex, then $\lambda(t)$ is superharmonic on B.

3. BOUNDARY BEHAVIOR OF THE ROBIN FUNCTION $\Lambda(\xi)$

We now consider a bounded domain $D \subset \mathbb{C}^n$ $(n \geq 2)$ with smooth boundary ∂D. Let $D' = D'_0 \cup D'_1 \cup \ldots \cup D'_q$ where each D'_i is a connected component of $\mathbb{C}^n - \bar{D}$ (there are only finitely many since ∂D is smooth) and D'_0 is the unbounded component. Given $\xi \in D$, we denote by $G(\xi, z)$ the Green function for (D, ξ) and by $\Lambda(\xi)$ the Robin constant, i.e.,

$$G(\xi, z) = \frac{1}{\|z-\xi\|^{2n-2}} + \Lambda(\xi) + H(\xi, z)$$

where $H(\xi, z)$ is harmonic for $z \in D$ with $H(\xi, \xi) = 0$. For $\xi \in D'_i$, these will denote the Green function and Robin constant for (D'_i, ξ). In section 8 [Y], it was shown that $\Lambda(\xi)$, as a function of $\xi \in D$, is a negative-valued real analytic function in D satisfying $\lim_{\xi \to \partial D} \Lambda(\xi) = -\infty$. In the next three chapters, we analyze the boundary behavior of the <u>Robin</u> <u>function</u> $\Lambda(\xi)$ more precisely.

Let $\psi(z)$ be a C^∞ function in all of $\mathbb{C}^n$ which defines D, i.e., $D = \{z \in \mathbb{C}^n: \psi(z) < 0\}$, $\partial D = \{z \in \mathbb{C}^n: \psi(z) = 0\}$ and Grad $\psi =$ $\left(\frac{\partial \psi}{\partial z_1}, \ldots, \frac{\partial \psi}{\partial z_n}\right) \neq (0, \cdots, 0)$ for all $z \in \partial D$. Without loss of generality we may assume that $\psi(z) \equiv c > 0$ for $\|z\|$ sufficiently large. For $\xi \in \mathbb{C}^n$, we define

$$(3.1) \qquad \lambda(\xi) \equiv \begin{cases} \Lambda(\xi)\psi(\xi)^{2n-2} & \text{if } \quad \xi \in \mathbb{C}^n - \partial D \\ -\|\mathrm{Grad}\ \psi(\xi)\|^{2n-2} & \text{if } \quad \xi \in \partial D \end{cases}$$

Our goal is to prove the following smoothness property of the function $\lambda(\xi)$.

<u>Lemma</u> <u>3.1</u>. $\lambda(\xi)$ is a C^2 function on $\mathbb{C}^n$.

We discuss the geometric meaning of the function $\lambda(\xi)$ in this chapter and then prove Lemma 3.1 in the next two chapters. Given $\xi \in D \cup D'$, we form the affine transformation

$$T_\xi(z) = w \equiv \frac{z-\xi}{-\psi(\xi)} .$$

Then for each $\xi \in \mathbb{C}^n$, we define

(3.2)

$$D(\xi) \equiv \begin{cases} T_\xi(D) & \text{if } \xi \in D \\ \left\{ w \in \mathbb{C}^n : 2\mathrm{Re}\left[\sum_{\alpha=1}^{n} \frac{\partial\psi}{\partial z_\alpha}(\xi) w_\alpha \right] - 1 < 0 \right\} & \text{if } \xi \in \partial D \\ T_\xi(D_i') & \text{if } \xi \in D_i' \ (i=0,1,\ldots,q) \end{cases}$$

and we set $\mathcal{D} \equiv \bigcup_{\xi\in\mathbb{C}^n} (\xi, D(\xi)) \subset \mathbb{C}^n \times \mathbb{C}^n$, $\partial\mathcal{D} \equiv \bigcup_{\xi\in\mathbb{C}^n} (\xi, \partial D(\xi))$. Thus we consider $\mathcal{D}$ as a variation of domains $D(\xi) \subset \mathbb{C}^n$ with parameter $\xi \in \mathbb{C}^n$, i.e., $\mathcal{D}$: $\xi \to D(\xi)$, $\xi \in \mathbb{C}^n$. The reason for the definition of $D(\xi)$ when $\xi \in \partial D$ will become apparent below.

Note that

(i) each $D(\xi)$ $(\xi \in \mathbb{C}^n)$ contains the origin;

(ii) for each $\xi \in D(D_i')$, $D(\xi)$ is similar to $D(D_i')$ with similarity ratio $\frac{1}{|\psi(\xi)|}$; by the smoothness of ∂D

(iii) $\mathcal{D} \cup \partial\mathcal{D}|_D$: $\xi \to \overline{D(\xi)}$, $\xi \in D$, is a variation which is diffeomorphically equivalent to the trivial variation $D \times \bar{D}$. Similarly, $\mathcal{D} \cup \partial\mathcal{D}|_{D_i'}$ is diffeomorphically equivalent to $D_i' \times \bar{D}_i'$, $i = 0,1,\cdots,q$.

We want to realize the variation $\mathcal{D}$ as being defined by a C^∞ function $f(\xi,w)$ in $\mathbb{C}^n \times \mathbb{C}^n$. To this end, we set, for $(\xi,w) \in \mathbb{C}^n \times \mathbb{C}^n$, $\tilde{\psi}(\xi,w) \equiv \psi(\xi-\psi(\xi)w)$. Then $\tilde{\psi} \in C^\infty(\mathbb{C}^n \times \mathbb{C}^n)$ and satisfies $\tilde{\psi}(\xi,w) = 0$ for $(\xi,w) \in \partial D \times \mathbb{C}^n$. On the other hand, if we consider $\psi(\xi) = \psi(\xi,w)$ as a function on $\mathbb{C}^n \times \mathbb{C}^n$ which is independent of $w \in \mathbb{C}^n$, then the smooth, real 4n-1 dimensional surface $\partial D \times \mathbb{C}^n$ is equal to $\{(\xi, w) \in \mathbb{C}^n \times \mathbb{C}^n : \psi(\xi, w) = 0\}$ with $\mathrm{Grad}_{(\xi)}\psi \neq 0$ on $\partial D \times \mathbb{C}^n$. Thus there exists a C^∞ function $f(\xi, w)$ on $\mathbb{C}^n \times \mathbb{C}^n$ satisfying $\tilde{\psi}(\xi, w) = -f(\xi, w)\psi(\xi)$. It follows that $f(\xi, 0) = -1$ for $\xi \in \mathbb{C}^n$ and

$$(3.3) \quad \frac{\partial f}{\partial w_\alpha}(\xi, w) = \frac{\partial \psi}{\partial z_\alpha}\Big|_{z=\xi-\psi(\xi)w} \quad \text{for } (\xi, w) \in \mathbb{C}^n \times \mathbb{C}^n, \ \alpha = 1, \cdots, n .$$

For each $\xi \in \mathbb{C}^n$, as a function of $w \in \mathbb{C}^n$ we can write

$$(3.4) \quad f(\xi, w) = \int_0^1 \left[\frac{d}{dt} f(\xi, tw)\right] dt + f(\xi, 0)$$

$$= \int_0^1 \sum_{\alpha=1}^n \left(w_\alpha \frac{\partial f}{\partial w_\alpha}(\xi, tw) + \bar{w}_\alpha \frac{\partial f}{\partial \bar{w}_\alpha}(\xi, tw)\right) dt - 1$$

$$= 2 \,\mathrm{Re} \left\{ \sum_{\alpha=1}^n \int_0^1 \left[w_\alpha \frac{\partial \psi}{\partial z_\alpha}\Big|_{z=\xi-\psi(\xi)tw}\right] dt \right\} - 1$$

by (3.3). In particular, if $\xi \in \partial D$ then $\psi(\xi) = 0$ so that

$$f(\xi, w) = 2 \,\mathrm{Re} \left\{ \sum_{\alpha=1}^n w_\alpha \frac{\partial \psi}{\partial z_\alpha}(\xi) \right\} - 1 .$$

It follows from the definition of $D(\xi)$ in (3.2) that $D(\xi) = \{w \in \mathbb{C}^n : f(\xi, w) < 0\}$ for all $\xi \in \mathbb{C}^n$, i.e., the variation $\mathcal{D}: \xi \to D(\xi)$, $\xi \in \mathbb{C}^n$ is defined by $f(\xi, w)$.

To discuss the geometric significance of the function $\lambda(\xi)$, fix $\xi \in \mathbb{C}^n$. By (i), each domain $D(\xi)$ has a Green function $g(\xi, w)$ with pole at 0 and Robin constant $\lambda(\xi)$ for $(D(\xi), 0)$, i.e.,

$$g(\xi, w) = \frac{1}{\|w\|^{2n-2}} + \lambda(\xi) + h(\xi, w)$$

where $h(\xi, w)$ is harmonic for $w \in D(\xi)$ and

(3.6) $$h(\xi,0) = 0 .$$

In particular, when $\xi \in \partial D$, $D(\xi)$ is a half-space and we have the explicit formulas

(3.7) $$g(\xi,w) = \frac{1}{\|w\|^{2n-2}} - \frac{1}{\|w-\bar{N}_\xi\|^{2n-2}}, \quad \lambda(\xi) = -\|\text{Grad}\ \psi(\xi)\|^{2n-2},$$

where $\bar{N}_\xi \equiv \overline{\text{Grad}\ \psi(\xi)}/\|\text{Grad}\ \psi(\xi)\|^2$ is the symmetric point of the origin 0 with respect to the hyperplane $\{w \in \mathbb{C}^n: \ f(\xi,w) = 0\}$. If $\xi \in D \cup D'$, since $T_\xi(z) = \frac{z-\xi}{-\psi(\xi)}$ is a similarity with similarity ratio $\frac{1}{|\psi(\xi)|}$, by Proposition 5.1 [Y] it follows that

(3.8) $$g(\xi,w) = G(\xi,z)\psi(\xi)^{2n-2} \quad \text{and} \quad \lambda(\xi) = \Lambda(\xi)\psi(\xi)^{2n-2}$$

Thus the function $\lambda(\xi)$, for $\xi \in \mathbb{C}^n$, is the Robin constant for $(D(\xi),0)$. Note that by the maximum principle

(3.9) $$0 < g(\xi,w) < \frac{1}{\|w\|^{2n-2}} .$$

As a specific example to make the above ideas concrete and also to motivate the content of Lemma 3.1, let $D = \{z \in \mathbb{C}^n: \ \|z\| < 1\}$, the unit ball in $\mathbb{C}^n$. We can take $\psi(z) = \|z\|^2 - 1$ for $z \in \mathbb{C}^n$. Since $D(\xi)$ is a ball for $\xi \in D$ and the complement of a ball if $\xi \in D'$, we easily compute that

$$\Lambda(\xi) = \frac{-1}{(1-\|\xi\|^2)^{2n-2}} \quad \text{if} \quad \xi \in D \cup D' .$$

Hence

$$\lambda(\xi) = \Lambda(\xi)\psi(\xi)^{2n-2} \equiv -1 \quad \text{for} \quad \xi \in D \cup D' .$$

If $\xi \in \partial D$, since $\text{Grad}\ \psi(\xi) = (\xi_1,\cdots,\xi_n)$, $\|\text{Grad}\ \psi(\xi)\| = 1$ so that

$\lambda(\xi) \equiv -1$ for each half-space $D(\xi) = \left\{w \in \mathbb{C}^n \colon\ 2 \operatorname{Re}\left(\sum_{\alpha=1}^{n} \xi_\alpha \bar{w}_\alpha\right) < 1\right\}$.

Thus in this simple example, the function $\lambda(\xi)$ is constant on $\mathbb{C}^n$. Note that $f(\xi, w) = -\psi(\xi)\ \|w\|^2 - 1 + 2 \operatorname{Re}\left(\sum_{\alpha=1}^{n} \xi_\alpha \bar{w}_\alpha\right)$.

4. PROOF OF LEMMA 3.1

In this chapter, we prove that $\lambda(\xi)$ is of class C^1 on $\mathbb{C}^n$. We first set up some notation. Let

$$\mathcal{D}^* \equiv \bigcup_{\xi\in D\cup D'} (\xi, D(\xi)), \quad \partial\mathcal{D}^* \equiv \bigcup_{\xi\in D\cup D'} (\xi, \partial D(\xi)).$$

<u>Step 1</u>. (1) $g(\xi,w)$ is of class C^4 for $(\xi,w) \in \mathcal{D}^* \cup \partial\mathcal{D}^* - (D \cup D' \times \{0\})$ and $\lambda(\xi)$ is of class C^4 on $D \cup D'$.

(2) $g(\xi,w)$ is continuous on $\mathcal{D} \cup \partial\mathcal{D} - \mathbb{C}^n \times \{0\}$

(3) $\dfrac{\partial g}{\partial w_\alpha}(\xi,w)$ is continuous on $\mathcal{D} \cup \partial\mathcal{D} - \mathbb{C}^n \times \{0\}$

(4) $\lambda(\xi)$ is continuous on $\mathbb{C}^n$.

<u>Proof of Step 1</u>. (1) follows from (iii) in Chapter 3. Thus to prove (2) we need only consider $\xi \in \partial D$. Fix $\xi_0 \in \partial D$ and $w_0 \in \overline{D(\xi_0)}$ and choose $R > \|w_0\|$. Since the function $f(\xi,w)$ which defines $\mathcal{D}$ is a C^∞ function on $\mathbb{C}^n \times \mathbb{C}^n$, we can find $r > 0$ sufficiently small so that if $b \equiv \{\xi \in \mathbb{C}^n\colon \|\xi - \xi_0\| < r\}$, then the variation

$$\mathcal{D}_{R,b}\colon \quad \xi \to \overline{D_R(\xi)}, \quad \xi \in b,$$

where $D_R(\xi) \equiv D(\xi) \cap \{w \in \mathbb{C}^n\colon \|w\| < R\}$, is diffeomorphically equivalent to the trivial variation $b \times \overline{D_R(\xi_0)}$. By Preliminary 4.2 [Y] it follows that the Green function $g_R(\xi,w)$ for $(D_R(\xi),0)$ is of class C^2 for $(\xi,w) \in \mathcal{D}_{R,b} \cup \partial\mathcal{D}_{R,b}$ except for $b \times \{0\}$ and the 'corners' $\bigcup_{\xi\in b} (\xi, e_R(\xi))$, where $e_R(\xi) \equiv \{w\colon \|w\| = R\} \cap \partial D(\xi)$.

Given $\xi \in b$, we consider the function $u_R(\xi,w) \equiv g(\xi,w) - g_R(\xi,w)$. Then $u_R(\xi,w)$ is harmonic for $w \in D_R(\xi)$ with boundary values

$$u_R(\xi,w) = \begin{cases} g(\xi,w) & \text{if } \|w\| = R,\ w \in D(\xi) \\ 0 & \text{if } \|w\| < R,\ w \in \partial D(\xi)\ . \end{cases}$$

By (3.9),

$$0 < g(\xi,w) < \frac{1}{\|w\|^{2n-2}} \quad \text{for} \quad w \in D(\xi)$$

so that $0 \le u_R(\xi,w) < \frac{1}{R^{2n-2}}$ for $w \in \partial D_R(\xi)$ and hence for all $w \in \overline{D_R(\xi)}$. Thus for $\xi \in b$ and $w \in \overline{D_R(\xi)}$,

$$\begin{aligned} |g(\xi,w) - g(\xi_0,w_0)| &\le |u_R(\xi,w) + g_R(\xi,w) - g_R(\xi_0,w_0) - u_R(\xi_0,w_0)| \\ &\le \frac{2}{R^{2n-2}} + |g_R(\xi,w) - g_R(\xi_0,w_0)|\ . \end{aligned}$$

Since R was an arbitrary number satisfying $R > \|w_0\|$ and $g_R(\xi,w)$ is continuous at (ξ_0,w_0) , it follows that $g(\xi,w)$ is continuous at (ξ_0,w_0) and (2) is proved. From Preliminary 4.2 and formula (1.3) [Y] , (3) and (4) follow. ∎

<u>Remark</u>. By similar reasoning used to show (3), it follows that $\frac{\partial g_a}{\partial w_\alpha}(\xi,w)$ is continuous in (a,ξ,w) where $a \in D(\xi)$ and $g_a(\xi,w)$ is the Green function for $(D(\xi),a)$. This fact will be used in Step 5.

<u>Step 2</u>. Fix $R > 0$ and let B be a ball of radius R. For $\xi \in \partial B$, let $E = E_\xi \equiv \{w : \|w - \xi\| < R\}$. Let Ω be a smoothly bounded domain in $\mathbb{C}^n - B$ such that $\partial\Omega$ is tangent to ∂B at ξ. Then for any harmonic function u in Ω satisfying

(a) $u(z) \le M$ for $z \in \Omega$ and

(b) $u(z) = 0$ for $z \in \partial\Omega \cap E$,

there exists a constant $c > 0$ which is independent of Ω and u such that

$$(4.1) \qquad \|\operatorname{Grad} u(\xi)\| \le \frac{cM}{R}.$$

Proof of Step 2. First suppose $M = R = 1$. We consider the harmonic function $w(z)$ on $E - \bar{B}$ with boundary values

$$w(z) = \begin{cases} 0 & \text{if } z \in \partial B \cap E \\ 1 & \text{if } z \in \partial E \cap (\mathbb{C}^n - \bar{B}) . \end{cases}$$

By the maximum principle, $u(z) \le w(z)$ on $\Omega \cap (E - \bar{B})$. Since $\partial\Omega$ and ∂B are tangent at ξ and $u(\xi) = w(\xi) = 0$, it follows that

$$\|\operatorname{Grad} u(\xi)\| = \frac{-1}{2}\frac{\partial u}{\partial n_z}(\xi) \le \frac{-1}{2}\frac{\partial w}{\partial n_z}(\xi) = \|\operatorname{Grad} w(\xi)\| \equiv c$$

where c does not depend on u or Ω. For the general case, the function $\tilde{u}(\tilde{z}) \equiv \frac{1}{M} u(z)$, where $\tilde{z} = \frac{1}{R} z$, satisfies (a) and (b) with $M = R = 1$; thus $\|\operatorname{Grad} \tilde{u}(\tilde{\xi})\| \le c$ so that $\|\operatorname{Grad} u(\xi)\| = M \dfrac{\|\operatorname{Grad} \tilde{u}(\tilde{\xi})\|}{R} \le \dfrac{cM}{R}$. ∎

Step 3. Let $D_1 = \{\xi \in \mathbb{C}^n : \operatorname{dist}(\xi, \partial D) < 1\}$. Then there exists a constant $K > 0$, independent of ξ and w, such that

$$(4.2) \qquad \|\operatorname{Grad}_{(w)} g(\xi)\| \le \frac{K}{\|w\|^{2n-1}}, \quad \forall\, \xi \in D_1, \quad w \in \partial D(\xi).$$

Proof of Step 3. Let $\ell = \sup \{\|\xi_1 - \xi_2\| : \xi_1, \xi_2 \in D\} + 2$. Since ∂D is smooth, there exist ρ, ρ' with $0 < \rho, \rho' < \ell$ such that for each $\xi \in \partial D$, we can find balls $b \subset D$ and $b' \subset D'$ of radii ρ and ρ' such that ∂b is internally tangent to D at ξ and $\partial b'$ is externally tangent to D at ξ. Let $\tilde{\rho} = \min(\rho, \rho')$.

We now use Step 2 to prove (4.2) if $\xi \in D_1 - \partial D$ and $w_0 \in \partial D(\xi)$. For such ξ, w_0, we can find a point $z_0 \in \partial D$ with $w_0 = \frac{z_0 - \xi}{-\psi(\xi)}$. Thus

$$\|w_0\| \le \frac{\ell}{|\psi(\xi)|}. \tag{4.3}$$

We remark for future use that inequality (4.3) holds for all $w_0 \in \overline{D(\xi)}$. Since $D(\xi)$ is similar to D (if $\xi \in D$) or D_i' (if $\xi \in D_i'$) with similarity ratio $\frac{1}{|\psi(\xi)|}$, there exists a ball $B \subset \mathbb{C}^n - D(\xi)$ with radius

$$\frac{\tilde{\rho}\|w_0\|}{\ell} < \frac{\tilde{\rho}}{|\psi(\xi)|}$$

(by (4.3)) such that ∂B is tangent to $D(\xi)$ at w_0. Let $E \equiv \{w : \|w - w_0\| < \frac{\tilde{\rho}\|w_0\|}{\ell}\}$. Then $w \in E$ implies that

$$\|w\| > \|w_0\| - \frac{\tilde{\rho}\|w_0\|}{\ell} = \left(1 - \frac{\tilde{\rho}}{\ell}\right)\|w_0\|.$$

Thus by (3.9),

$$0 < g(\xi, w) < \frac{1}{\|w\|^{2n-2}} < \left(\frac{1}{1 - \frac{\tilde{\rho}}{\ell}}\right)^{2n-2} \frac{1}{\|w_0\|^{2n-2}} \quad \text{for } w \in E \cap D(\xi).$$

By Step 2, since $g(\xi, w)$ is harmonic in $E \cap D(\xi)$ (note $\tilde{\rho} < \ell$ implies that $0 \notin E$) and $g(\xi, w) = 0$ for $w \in E \cap \partial D(\xi)$,

$$\|\operatorname{Grad}_{(w)} g(\xi, w)\| \le c\left(\frac{1}{1 - \frac{\tilde{\rho}}{\ell}}\right)^{2n-2} \frac{1}{\|w_0\|^{2n-2}} \Big/ \frac{\tilde{\rho}\|w_0\|}{\ell} \equiv K_1 \frac{1}{\|w_0\|^{2n-1}} \tag{4.4}$$

where $K_1 = \dfrac{c\ell}{\tilde{\rho}\left(1 - \frac{\tilde{\rho}}{\ell}\right)^{2n-2}}$ is independent of $\xi \in D_1 - \partial D$ and $w_0 \in \partial D(\xi)$.

By Step 1(3), or by explicit computation of $\|\operatorname{Grad}_{(w)} g(\xi, w_0)\|$ from (3.2), inequality (4.4), with perhaps a different constant K_1, remains valid for $\xi \in \partial D$ and $w_0 \in \partial D(\xi)$. Thus (4.2) is proved. ∎

Before proceeding to Step 4, we note that since $0 \in D(\xi)$ for all $\xi \in \mathbb{C}^n$ and $f(\xi,w)$ defines the smooth variation D, by Theorem 2.1 (or its proof) we have, for $\xi \in D \cup D'$, $a \in D(\xi)$, $\nu = 1,\cdots,n$,

$$(4.5)\quad \frac{\partial g}{\partial \xi_\nu}(\xi,a) = \frac{1}{2(n-1)w_{2n}} \int_{\partial D(\xi)} K_1^{(\nu)}(\xi,w)\|\mathrm{Grad}_{(w)} g(\xi,w)\| \frac{\partial g_a(\xi,w)}{\partial n_w}\, dS_w$$

$$(4.6)\quad \frac{\partial^2 g}{\partial \xi_\nu \partial \bar{\xi}_\nu}(\xi,a) = \frac{1}{2(n-1)w_{2n}} \int_{\partial D(\xi)} K_2^{(\nu)}(\xi,w)\|\mathrm{Grad}_{(w)} g(\xi,w)\| \frac{\partial g_a(\xi,w)}{\partial n_w}\, dS_w$$

$$- \frac{4}{(n-1)w_{2n}} \mathrm{Re} \iint_{D(\xi)} \left[\sum_{\alpha=1}^{n} \frac{\partial^2 g(\xi,w)}{\partial \xi_\nu \partial \bar{w}_\alpha} \frac{\partial^2 g_a(\xi,w)}{\partial \bar{\xi}_\nu \partial w_\alpha}\right] dV_w \equiv I_g^{(\nu)}(\xi,a) + J_g^{(\nu)}(\xi,a)$$

where

$$(4.7)\qquad K_1^{(\nu)}(\xi,w) \equiv \frac{\partial f}{\partial \xi_\nu}(\xi,w)/\|\mathrm{Grad}_{(w)} f(\xi,w)\|$$

and

$$(4.8)\quad K_2^{(\nu)}(\xi,w) \equiv \frac{\mathcal{L}^{(\nu)} f}{\|\mathrm{Grad}_{(w)} f\|^3}$$

$$\equiv \frac{1}{\|\mathrm{Grad}_{(w)} f\|^3}\left[\frac{\partial^2 f}{\partial \xi_\nu \partial \bar{\xi}_\nu}\|\mathrm{Grad}_{(w)} f\|^2 - 2\,\mathrm{Re}\left\{\sum_{\beta=1}^{n} \frac{\partial f}{\partial \xi_\nu} \frac{\partial f}{\partial \bar{w}_\beta} \frac{\partial^2 f}{\partial w_\beta \partial \bar{\xi}_\nu}\right\} + \left|\frac{\partial f}{\partial \bar{\xi}_\nu}\right|^2 \Delta_{(w)} f\right]$$

for $w \in \partial D(\xi)$, $\nu = 1,\cdots,n$. For simplicity in notation, we fix , $\nu \in \{1,2,\cdots,n\}$ and denote $\mathcal{L}^{(\nu)}$, $K_1^{(\nu)}$, $K_2^{(\nu)}$, $I_g^{(\nu)}$, $J_g^{(\nu)}$, by $\mathcal{L}$, K_1, K_2, I_g, J_g. We want to show that for $\xi_0 \in \partial D$ and $a_0 \in D(\xi_0)$

$$(4.9)\quad \lim_{\substack{\xi\to\xi_0 \\ a\to a_0}} \frac{\partial g}{\partial \xi_\nu}(\xi,a) = \frac{1}{2(n-1)w_{2n}} \int_{\partial D(\xi_0)} K_1(\xi_0,w)\|\mathrm{Grad}_{(w)} g(\xi_0,w)\| \frac{\partial g_a(\xi_0,w)}{\partial n_w}\, dS_w$$

and

$$(4.10)\quad \lim_{\substack{\xi\to\xi_0\\ a\to a_0}} \frac{\partial^2 g}{\partial\xi_\nu\partial\bar{\xi}_\nu}(\xi,a) = I_g(\xi_0,a_0) + J_g(\xi_0,a_0) .$$

Then a standard argument, together with Step 1, will show that (4.9) implies $g(\xi,w)$ is C^1 in $\mathcal{D} - \mathbb{C}^n \times \{0\}$ so that $\lambda(\xi)$ is C^1 in $\mathbb{C}^n$. Equation (4.10) will be proved in Chapter 5 and will be used to show that $\lambda(\xi)$ is actually of class C^2 in $\mathbb{C}^n$. To prove (4.9) and (4.10) we need to estimate derivatives of f. Note first that

$$(3.3)\quad \frac{\partial f}{\partial w_\alpha}(\xi,w) = \frac{\partial\psi}{\partial z_\alpha}\Big|_{z=\xi-\psi(\xi)w}$$

implies that

$$(4.11)\quad \min_{\substack{\xi\in\partial D\\ w\in D(\xi)}} \|\mathrm{Grad}_{(w)} f(\xi,w)\| = \min_{\xi\in\partial D} \|\mathrm{Grad}_{(w)}\psi(\xi)\| \equiv m > 0.$$

<u>Step</u> <u>4</u>. There exists a constant $A > 0$ such that for all $\xi \in D_1 \equiv \{\xi \in \mathbb{C}^n : \mathrm{dist}(\xi,\partial D) < 1\}$ and $w \in \overline{D(\xi)}$ with $\|w\| > 1$,

(1) $|f(\xi,w)| \le A\|w\|$

(2) $\left|\frac{\partial f}{\partial w_\alpha}(\xi,w)\right| \le A$

(3) $\left|\frac{\partial f}{\partial \xi_\nu}(\xi,w)\right| \le A\|w\|^2$

(4) $\left|\frac{\partial^2 f}{\partial w_\alpha\partial w_\beta}(\xi,w)\right| \le \frac{A}{\|w\|}$

(5) $\left|\frac{\partial^2 f}{\partial \xi_\nu\partial w_\alpha}(\xi,w)\right| \le A\|w\|$

(6) $\left|\frac{\partial^2 f}{\partial \xi_\nu\partial \xi_\mu}(\xi,w)\right| \le A\|w\|^3$.

<u>Proof</u> <u>of</u> <u>Step</u> <u>4</u>. First of all, since we can assume $\psi(z) \equiv c > 0$ for $\|z\|$ large, we can find $M > 0$ so that

$$|\psi|,\ \left|\frac{\partial\psi}{\partial z_\alpha}\right|,\ \left|\frac{\partial^2\psi}{\partial z_\alpha\partial z_\beta}\right|,\ \left|\frac{\partial^2\psi}{\partial z_\alpha\partial\bar{z}_\beta}\right| < M \text{ on } \mathbb{C}^n, \quad \alpha,\beta = 1,\cdots,n.$$

From equation

$$(3.3) \qquad \frac{\partial f}{\partial w_\alpha}(\xi, w) = \frac{\partial \psi}{\partial z_\alpha}\Big|_{z=\xi-\psi(\xi)w} .$$

we recall that equation (3.4) may be written as

$$(3.4) \qquad f(\xi, w) = \int_0^1 \sum_{\beta=1}^{n} \left(w_\beta \frac{\partial \psi}{\partial z_\beta} + \bar{w}_\beta \frac{\partial \psi}{\partial \bar{z}_\beta} \right) \Big|_{z=\xi-\psi(\xi)tw} dt - 1.$$

Hence

$$(1)' \qquad |f(\xi, w)| \le 2\sqrt{n}\, M\|w\| + 1 \quad \text{for} \quad (\xi, w) \in \mathbb{C}^n \times \mathbb{C}^n$$

which yields (1) and

$$(2)' \qquad \left|\frac{\partial f}{\partial w_\alpha}(\xi, w)\right| \le M \quad \text{for} \quad (\xi, w) \in \mathbb{C}^n \times \mathbb{C}^n$$

which gives (2).

Differentiating (3.4) with respect to ξ_ν, we obtain

$$(4.12) \qquad \frac{\partial f}{\partial \xi_\nu}(\xi, w) = \sum_{\beta=1}^{n} \int_0^1 \left(w_\beta \frac{\partial^2 \psi}{\partial z_\nu \partial z_\beta} + \bar{w}_\beta \frac{\partial^2 \psi}{\partial z_\nu \partial \bar{z}_\beta} \right) \Big|_{z=\xi-\psi(\xi)tw} dt$$

$$-2 \frac{\partial \psi}{\partial \xi_\nu}(\xi) \operatorname{Re}\left[\sum_{\beta,\gamma=1}^{n} \int_0^1 \left[w_\beta w_\gamma \frac{\partial^2 \psi}{\partial z_\beta \partial z_\gamma} + \bar{w}_\beta w_\gamma \frac{\partial^2 \psi}{\partial \bar{z}_\beta \partial z_\gamma} \right] \Big|_{z=\xi-\psi(\xi)tw} t\, dt \right]$$

so that

$$(3)' \qquad \left|\frac{\partial f}{\partial \xi_\nu}(\xi, w)\right| \le 2\sqrt{n}\, M\|w\| + 4n\, M^2\|w\|^2 \quad \text{for} \quad (\xi, w) \in \mathbb{C}^n \times \mathbb{C}^n$$

which gives (3). Differentiating (3.3) with respect to w_β yields

$$(4.13) \qquad \frac{\partial^2 f}{\partial w_\beta \partial w_\alpha} = \frac{\partial^2 \psi}{\partial z_\beta \partial z_\alpha}(-\psi(\xi))$$

so that

$$\left|\frac{\partial^2 f}{\partial w_\beta \partial w_\alpha}(\xi, w)\right| \le M \quad \text{for} \quad (\xi, w) \in \mathbb{C}^n \times \mathbb{C}^n.$$

We need a better estimate to obtain (4). We recall that by the remark after equation (4.3), $|\psi(\xi)| \le \frac{\ell}{\|w\|}$ for $\xi \in D_1$, $w \in \overline{D(\xi)} - \{0\}$. Hence (4.13), together with this above inequality, imply that

$$(4)' \qquad \left|\frac{\partial^2 f}{\partial w_\beta \partial w_\alpha}(\xi, w)\right| \le M \frac{\ell}{\|w\|} \quad \text{for} \quad \xi \in D_1, \; w \in \overline{D(\xi)} - \{0\}$$

which gives (4). Finally, differentiating (4.12) with respect to w_α (resp. ξ_μ) and estimating in the obvious way yields (5) (resp. (6)). ∎

<u>Remark</u>. It is clear that the estimates (1)-(3), (5), and (6) (if we add a constant) hold for <u>all</u> $w \in \overline{D(\xi)}$; also, higher order derivatives of f can be estimated in a similar fashion.

Note that since $\psi(\xi_0) = 0$ for $\xi_0 \in \partial D$, by (4.12) we have

$$(4.14) \qquad \frac{\partial f}{\partial \xi_\nu}(\xi_0, w) = \sum_{\beta=1}^{n} \left[\frac{\partial^2 \psi}{\partial z_\nu \partial z_\beta}(\xi_0) w_\beta + \frac{\partial^2 \psi}{\partial z_\nu \partial \bar{z}_\beta}(\xi_0) \bar{w}_\beta\right]$$

$$- \frac{\partial \psi}{\partial \xi_\nu}(\xi_0) \operatorname{Re}\left[\sum_{\beta,\gamma=1}^{n} \left(\frac{\partial^2 \psi}{\partial z_\beta \partial z_\gamma}(\xi_0) w_\beta w_\gamma + \frac{\partial^2 \psi}{\partial \bar{z}_\beta \partial z_\gamma}(\xi_0) \bar{w}_\beta w_\gamma\right)\right], \quad \xi_0 \in \partial D,$$

which is a function of degree 2 in $w_\alpha, \bar{w}_\beta$. This formula will be used in Chapter 11.

<u>Step 5</u>. (4.9) holds.

<u>Proof of Step 5</u>. Fix $R > 1$. If $\xi \in D \cup D'$ approaches $\xi_0 \in \partial D$, then the boundary surfaces $\partial D(\xi) \cap \{w : \|w\| < R\}$ approach $\partial D(\xi_0) \cap \{w : \|w\| < R\}$ continuously in the sense that the unit normal vectors

$$\frac{\operatorname{Grad}_{(w)} g(\xi, w)}{\|\operatorname{Grad}_{(w)} g(\xi, w)\|}$$

converge uniformly on compact sets to

$$\frac{\mathrm{Grad}_{(w)} g(\xi_0, w)}{\|\mathrm{Grad}_{(w)} g(\xi_0, w)\|}$$

except at the corners $\partial D(\xi) \cap \{w : \|w\| = R\}$. By the remark after Step 1, $\dfrac{\partial g_a}{\partial w_\alpha}(\xi, w)$ is continuous so that

$$(4.15) \qquad \lim_{\substack{\xi\to\xi_0 \\ a\to a_0}} \int_{\partial D(\xi) \cap \{w : \|w\| < R\}} K_1(\xi, w)\|\mathrm{Grad}_{(w)} g(\xi, w)\| \frac{\partial g_a(\xi, w)}{\partial n_w} dS_w$$

$$= \int_{\partial D(\xi_0) \cap \{w : \|w\| < R\}} K_1(\xi_0, w)\|\mathrm{Grad}_{(w)} g(\xi, w)\| \frac{\partial g_a(\xi_0, w)}{\partial n_w} dS_w.$$

Hence it suffices to show that for $\xi \in D_1$ and $a \in D(\xi)$,

$$(4.16) \qquad \left| \int_{\partial D(\xi) \cap \{w : \|w\| > R\}} K_1(\xi, w)\|\mathrm{Grad}_{(w)} g(\xi, w)\| \frac{\partial g_a(\xi, w)}{\partial n_w} dS_w \right| \le O\left(\frac{1}{R}\right).$$

To prove (4.16), we use our estimates from Steps 3 and 4. By (4.11) and (3) of Step 4, for $\xi \in D_1$ and $\|w\| > 1$,

$$|K_1(\xi, w)| = \frac{\left|\frac{\partial f}{\partial \xi_\alpha}(\xi, w)\right|}{\|\mathrm{Grad}_{(w)} f(\xi, w)\|} \le \frac{A\|w\|^2}{m}.$$

Using this estimate together with (4.2) in (4.16), we obtain

$$\left| \int_{\partial D(\xi) \cap \{w : \|w\| > R\}} K_1(\xi, w)\|\mathrm{Grad}_{(w)} g(\xi, w)\| \frac{\partial g_a(\xi, w)}{\partial n_w} dS_w \right|$$

$$\le \frac{KA}{m} \frac{1}{R^{2n-3}} \int_{\partial D(\xi)} \left(\frac{-\partial g_a(\xi, w)}{\partial n_w} \right) dS_w$$

for $\xi \in D_1$ and $a \in D(\xi)$. Since $\int_{\partial D(\xi)} \left(\frac{-\partial g_a(\xi,w)}{\partial n_w}\right) dS_w = (2n-2)w_{2n}$,

(4.16) follows since $n \geq 2$. ∎

We remark that a rather explicit computation of the right-hand side of (4.9) will be carried out in Chapter 11 with the aid of equation (4.14).

Step 6. (1) $g(\xi,w)$ is of class C^1 on $\mathcal{D} - \mathbb{C}^n \times \{0\}$.

(2) $\lambda(\xi)$ is of class C^1 on $\mathbb{C}^n$.

Proof of Step 6. By Step 1, all that remains is to prove these assertions for $\xi \in \partial D$. As previously noted, (2) follows from (1) (cf. formula (1.3) [Y]). Fix $\xi_0 \in \partial D$ and $a_0 \in D(\xi_0) - \{0\}$. Choose $\rho, r > 0$ sufficiently small so that

$$B \times U \equiv \{(\xi,a) : \|\xi - \xi_0\| < \rho, \ \ \|a - a_0\| < r\} \subset \mathcal{D} - B \times \{0\}.$$

By Step 5,

(4.17)

$$g_\alpha(\xi,a) \equiv \begin{cases} \dfrac{\partial g}{\partial \xi_\alpha}(\xi,w) & \text{if } \xi \in B - \partial D \\ \dfrac{1}{2(n-1)w_{2n}} \displaystyle\int_{\partial D(\xi)} K_1(\xi,w)\|\mathrm{Grad}_{(w)} g(\xi,w)\| \frac{\partial g_a}{\partial n_w}(\xi,w)dS_w & \text{if } \xi \in B \cap \partial D \end{cases}$$

is a continuous function for $(\xi,a) \in B \times U$. We show that $\frac{\partial g}{\partial \xi_\alpha}(\xi,a)$ exists at ξ_0 and equals $g_\alpha(\xi_0,a)$. This proves that $\frac{\partial g}{\partial \xi_\alpha}(\xi,w)$ is continuous on $\mathcal{D} - \mathbb{C}^n \times \{0\}$; since g is real-valued, $\frac{\partial g}{\partial \bar{\xi}_\alpha} = \overline{\frac{\partial g}{\partial \xi_\alpha}}$ and (1) follows.

Fix a point $\xi^x \in B - \partial D$ and $a \in U$. Suppose, first of all, that $\xi^x \in D$. Consider the line integral

$$F(\xi) \equiv \int_{C(\xi^X,\xi)} \sum_{\alpha=1}^{n} [g_\alpha(\xi,a)d\xi_\alpha + \overline{g_\alpha(\xi,a)d\xi_\alpha}]$$

taken over a curve $C(\xi^X,\xi) \subset B$ connecting ξ^X and ξ. We show that

$$(4.18) \qquad F(\xi) = g(\xi,a) - g(\xi^X,a);$$

in particular, this implies that the line integral is independent of the curve $C(\xi^X,\xi)$ joining ξ^X and ξ. Let $\xi_1,\cdots,\xi_q$ be the intersection points of C and ∂D listed in order of occurence from ξ^X to ξ. If we write

$$C = C_0 \cup C_1 \cup \cdots \cup C_q$$

where C_i is the subarc of C joining ξ_i and ξ_{i+1} ($\xi_0 = \xi^X$ and $\xi_{q+1} = \xi$), then C_i lies entirely in $B - \partial D$ except for the endpoints. Since g and $\frac{\partial g}{\partial \xi_\alpha}$ are continuous up to ∂D with respect to ξ, it follows from (4.17) that for $\xi', \xi'' \in C_i$,

$$\int_{C_i} \sum_{\alpha=1}^{n} [g_\alpha(\xi,a)d\xi_\alpha + \overline{g_\alpha(\xi,a)d\xi_\alpha}]$$

$$= \lim_{\substack{\xi'\to\xi_i \\ \xi''\to\xi_{i+1}}} \int_{\xi'}^{\xi''} dg(\xi,a) = g(\xi_{i+1},a) - g(\xi_i,a),$$

which proves (4.18).

Since $F(\xi)$ is differentiable on B, it follows that $\frac{\partial g}{\partial \xi_\alpha}$ exists, even at ξ_0, and equals $\frac{\partial F}{\partial \xi_\alpha} = g_\alpha$, as desired. ∎

<u>Remark</u>. Using equation (4.17) and our estimates in Step 4, we have proven the following result.

<u>Corollary</u> <u>4.1.</u> For each $\xi \in D_1$, $\frac{\partial g}{\partial \xi_\alpha}(\xi, w)$ is a harmonic function of $w \in D(\xi)$ with boundary values

$$F(\xi, w) \equiv \frac{-\frac{\partial f}{\partial \xi_\alpha}(\xi, w)}{\|\mathrm{Grad}_{(w)} f(\xi, w)\|} \|\mathrm{Grad}_{(w)} g(\xi, w)\| \tag{4.19}$$

which satisfy

$$|F(\xi, w)| \le \frac{A}{\|w\|^{2n-3}} \quad \text{for} \quad w \in \partial D(\xi), \quad \xi \in D_1 \tag{4.20}$$

for some $A > 0$ which is independent of ξ, w.

5. PROOF OF LEMMA 3.1, CONTINUED

We now use some refinements of the estimates in Chapter 4 to prove that $\lambda(\xi)$ is of class C^2 in $\mathbb{C}^n$. The main ingredient is the continuity of $\frac{\partial^2 g}{\partial\xi_\nu\partial\bar{\xi}_\nu}(\xi,w)$ in $\mathcal{D}$, $\nu = 1,2,\cdots,n$. Fixing $\nu \in \{1,\cdots,n\}$, we have, recalling formula (4.6) for $\frac{\partial^2 g}{\partial\xi_\nu\partial\bar{\xi}_\nu}(\xi,a)$ and setting $a = 0$ to simplify the notation

$$(5.1)\quad \frac{\partial^2\lambda}{\partial\xi_\nu\partial\bar{\xi}_\nu}(\xi,w) = \frac{1}{2(n-1)w_{2n}} \int_{\partial D(\xi)} K_2(\xi,w)\ \|\mathrm{Grad}_{(w)} g(\xi,w)\|\ \frac{\partial g}{\partial n_w}(\xi,w)dS_w$$

$$- \frac{4}{(n-1)w_{2n}} \iint_{D(\xi)} \sum_{\alpha=1}^{n} \left|\frac{\partial^2 g}{\partial\xi_\nu\partial\bar{w}_\alpha}(\xi,w)\right|^2 dV_w \equiv I_g(\xi,0) + J_g(\xi,0)$$

where

$$K_2(\xi,w) = \frac{1}{\|\mathrm{Grad}_{(w)}f\|^3}\left[\frac{\partial^2 f}{\partial\xi_\nu\partial\bar{\xi}_\nu}\ \|\mathrm{Grad}_{(w)}f\|^2 - 2\,\mathrm{Re}\left\{\sum_{\beta=1}^{n}\frac{\partial f}{\partial\xi_\nu}\,\frac{\partial f}{\partial\bar{w}_\beta}\,\frac{\partial^2 f}{\partial w_\beta\partial\bar{\xi}_\nu}\right\}\right.$$

$$(5.2)\qquad \left. + \left|\frac{\partial f}{\partial\xi_\nu}\right|^2 \Delta_{(w)}f\right]$$

for $w \in \partial D(\xi)$. Our goal is to prove

$$(5.3)\qquad \lim_{\xi\to\xi_0}\,[I_g(\xi,0) + J_g(\xi,0)] = I_g(\xi_0,0) + J_g(\xi_0,0)$$

where $\xi \in D \cup D'$ and $\xi_0 \in \partial D$. We do this in two stages:

$$(\mathrm{I})\qquad \lim_{\xi\to\xi_0} I_g(\xi,0) = I_g(\xi_0,0)$$

and

$$(\mathrm{II})\qquad \lim_{\xi\to\xi_0} J_g(\xi,0) = J_g(\xi_0,0).$$

Step 1. (I) holds.

Proof of Step 1. For each fixed $\xi \in \mathbb{C}^n$, $\|\mathrm{Grad}_{(w)} f(\xi, w)\| \neq 0$ for $w \in \partial D(\xi)$. Hence $K_2(\xi, w)$ is clearly continuous in (ξ, w) for w near $\partial D(\xi)$; thus by the same reasoning as in Chapter 4, step 5, for $\xi_0 \in \partial D$ and $R > 1$,

$$\lim_{\xi \to \xi_0} \left[\int_{\partial D(\xi) \cap \{w:\, \|w\| < R\}} K_2(\xi, w)\, \|\mathrm{Grad}_{(w)} g(\xi, w)\| \frac{\partial g}{\partial n_w}(\xi, w)\, dS_w \right]$$

$$= \int_{\partial D(\xi_0) \cap \{w:\, \|w\| < R\}} K_2(\xi_0, w)\, \|\mathrm{Grad}_{(w)} g(\xi_0, w)\| \frac{\partial g}{\partial n_w}(\xi_0, w)\, dS_w .$$

Thus to prove (I) it suffices to prove that given $\varepsilon > 0$, we can find $r = r(\varepsilon)$ and $R = R(\varepsilon)$ such that for each $\xi \in U \equiv \{\xi : \|\xi - \xi_0\| < r\}$,

$$\left| \int_{\partial D(\xi) \cap \{w:\, \|w\| > R\}} K_2(\xi, w)\, \|\mathrm{Grad}_{(w)} g(\xi, w)\| \frac{\partial g}{\partial n_w}(\xi, w)\, dS_w \right| < \varepsilon \tag{5.4}$$

We may assume $\xi \in D_1$ in proving (5.4) (i.e., $r < 1$). From (5.2) and the estimates in Steps 3 and 4 of Chapter 4,

$$|K_2(\xi, \omega)| \le \frac{1}{m^3} [A\|w\|^3 A^2 + 2n(A\|w\|^2\, AA\|w\|) + (A\|w\|^2)^2 \frac{nA}{\|w\|}] \equiv B\|w\|^3$$

if $w \in \partial D(\xi)$ with $\|w\| > 1$. Thus if $R > 1$,

$$\left| \int_{\partial D(\xi) \cap \{w:\, \|w\| > R\}} K_2(\xi, w)\, \|\mathrm{Grad}_{(w)} g(\xi, w)\| \frac{\partial g}{\partial n_w}(\xi, w)\, dS_w \right|$$

$$\le \int_{\partial D(\xi) \cap \{w:\, \|w\| > R\}} B\|w\|^3 \frac{K}{\|w\|^{2n-1}} \left(\frac{-\partial g}{\partial n_w}(\xi, w) \right) dS_w$$

$$\le \frac{BK}{R^{2n-4}} \int_{\partial D(\xi) \cap \{w:\, \|w\| > R\}} \left(\frac{-\partial g}{\partial n_w}(\xi, w) \right) dS_w .$$

Since $\int_{\partial D(\xi)} \left(\frac{-\partial g}{\partial n_w}(\xi, w)\right) dS_w = (2n-2)w_{2n}$ for <u>all</u> $\xi \in \mathbb{C}^n$, given $\epsilon > 0$, we can find $R = R(\varepsilon)$ and $r = r(\varepsilon)$ so that

$$\int_{\partial D(\xi) \cap \{w: \|w\| > R\}} \left(\frac{-\partial g}{\partial n_w}(\xi, w)\right) dS_w < \varepsilon$$

for all ξ satisfying $\|\xi - \xi_0\| < r$. Since $n \geq 2$, choosing such R, r yields (5.4.) ∎

The proof of (II) is much more difficult and will require several steps. Note first of all that by Stokes' theorem it suffices to prove

$$\text{(II)}' \lim_{\xi \to \xi_0} \int_{\partial D(\xi)} \frac{\partial g}{\partial \xi_\nu} \frac{\partial^2 g}{\partial \bar{\xi}_\nu \partial w_\alpha} dw_\alpha \wedge dw_1 \wedge d\bar{w}_1 \wedge \cdots \widehat{dw_\alpha \wedge d\bar{w}_\alpha} \cdots \wedge dw_n \wedge d\bar{w}_n$$

$$= \int_{\partial D(\xi_0)} \frac{\partial g}{\partial \xi_\nu} \frac{\partial^2 g}{\partial \bar{\xi}_\nu \partial w_\alpha} dw_\alpha \wedge dw_1 \wedge d\bar{w}_1 \wedge \cdots \widehat{dw_\alpha \wedge d\bar{w}_\alpha} \cdots \wedge dw_n \wedge d\bar{w}_n$$

for $\alpha = 1, \cdots, n$. Clearly we need to study the behavior of $\frac{\partial g}{\partial \xi_\nu}(\xi, w)$ and $\frac{\partial^2 g}{\partial \bar{\xi}_\nu \partial w_\alpha}(\xi, w)$ for $w \in \partial D(\xi)$ as $\xi \to \xi_0 \in \partial D$.

Recalling Corollary 4.1, $\frac{\partial g}{\partial \xi_\nu}(\xi, w)$ is, for each $\xi \in D_1$, a harmonic function of $w \in D(\xi)$ with boundary values

$$(4.19) \qquad F(\xi, w) \equiv \frac{\frac{-\partial f}{\partial \xi_\nu}(\xi, w)}{\|\mathrm{Grad}_{(w)} f(\xi, w)\|} \|\mathrm{Grad}_{(w)} g(\xi, w)\| = \frac{\frac{1}{2} \frac{\partial f}{\partial \xi_\nu} \frac{\partial g}{\partial n_w}}{\|\mathrm{Grad}_{(w)} f\|}$$

which satisfy

(4.20) $$|F(\xi,w)| \le \frac{A}{\|w\|^{2n-3}} \quad \text{for} \quad w \in \partial D(\xi)$$

where $A > 0$ is a constant independent of ξ and w. Given a domain $\Omega \subset \mathbb{C}^n$ with smooth boundary and given a continuous function h on $\partial\Omega$, we introduce the notation $H_h = H_h^{\Omega}$ for the harmonic function in Ω with boundary values h. Thus we write $\frac{\partial g}{\partial \xi_\nu}(\xi,w) = H_F^{D(\xi)}(w)$ (or $\frac{\partial g}{\partial \xi_\nu} = H_F$ for short). The idea behind the proof of (II)′ is, as usual, to show first of all that for each $R > 1$,

(A)

$$\lim_{\xi\to\xi_0} \int_{\partial D(\xi)\cap\{w:\,\|w\|<R\}} \frac{\partial g}{\partial \xi_\nu}\,\frac{\partial^2 g}{\partial \bar\xi_\nu \partial w_\alpha}\, dw_\alpha \wedge dw_1 \wedge d\bar w_1 \wedge\cdots \widehat{dw_\alpha \wedge d\bar w_\alpha} \cdots\wedge dw_n \wedge d\bar w_n$$

$$= \int_{\partial D(\xi_0)\cap\{w:\,\|w\|<R\}} \frac{\partial g}{\partial \xi_\nu}\,\frac{\partial^2 g}{\partial \bar\xi_\nu \partial w_\alpha}\, dw_\alpha \wedge dw_1 \wedge d\bar w_1 \wedge\cdots \widehat{dw_\alpha \wedge d\bar w_\alpha} \cdots\wedge dw_n \wedge d\bar w_n$$

Using complex notation for dS_w, for $w \in \partial D(\xi)$ we have

(5.5) $$\frac{\partial g}{\partial n_w}\, dS_w = \frac{-i^n}{2^{n-1}} \sum_{\alpha=1}^{n} \frac{\partial g}{\partial w_\alpha}\, dw_\alpha \wedge dw_1 \wedge d\bar w_1 \wedge\cdots \widehat{dw_\alpha \wedge d\bar w_\alpha} \cdots\wedge dw_n \wedge d\bar w_n$$

(this follows since $g(\xi,w) = 0$ for $w \in \partial D(\xi)$; cf. [Y], Proposition 3.3). Thus using (4.19) we must also show that

(B)

$$\lim_{R\to+\infty} \int_{\partial D(\xi)\cap\{w:\,\|w\|>R\}} \frac{\frac{\partial f}{\partial \xi_\nu}}{\|\mathrm{Grad}_{(w)} f\|}\, \frac{\partial^2 g}{\partial \bar\xi_\nu \partial w_\alpha}\, \frac{\partial g}{\partial n_w}\, dS_w = 0$$

uniformly for ξ in a neighborhood of $\xi_0 \in \mathbb{C}^n$.

The proof of (A) requires showing that $\frac{\partial g}{\partial \xi_\nu}$ and $\frac{\partial^2 g}{\partial \bar{\xi}_\nu \partial w_\alpha}$ are continuous at $(\xi_0, w_0) \in \partial D \times \partial D(\xi_0)$, while the proof of (B) requires estimates on $\frac{\partial^2 g}{\partial \bar{\xi}_\nu \partial w_\alpha}(\xi, w)$ for $\|w\|$ large. Since $\frac{\partial g}{\partial \xi_\nu} = H_F$, the key ingredient is a precise estimate on F, $\frac{\partial F}{\partial w_\alpha}$, and $\frac{\partial^2 F}{\partial w_\alpha \partial \bar{w}_\alpha}$ for ξ near ξ_0 and w near w_0. Then a standard potential theory lemma will yield the desired properties of $\frac{\partial g}{\partial \xi_\nu}$, $\frac{\partial^2 g}{\partial \bar{\xi}_\nu \partial w_\alpha}$. The technical details are rather lengthy; we first state and prove the above-mentioned potential theory lemma (cf., [PW]).

Lemma 5.1. Let Ω be a bounded domain in $\mathbb{C}^n$ $(n \geq 2)$ with smooth boundary and let $\phi \in C^2(\bar{\Omega})$. Then there exists a constant $C > 0$ depending only on Ω and n such that

$$(1) \quad \frac{\partial \phi}{\partial n_w}(w) - C \max_{w \in \bar{\Omega}} |\Delta_{(w)} \phi(w)| \leq \frac{\partial H_\phi^\Omega}{\partial n_w}(w) \leq \frac{\partial \phi}{\partial n_w}(w) + C \max_{w \in \bar{\Omega}} |\Delta_{(w)} \phi(w)|$$

(5.6) and

$$(2) \quad \|\mathrm{Grad}\, H_\phi^\Omega(w)\| \leq \|\mathrm{Grad}\, \phi(w)\| + C \max_{w \in \bar{\Omega}} |\Delta_{(w)} \phi(w)|$$

for all $w \in \partial\Omega$.

Proof. Since Ω is bounded and $\partial\Omega$ is smooth, we can find $\rho > 0$ such that at each $w_0 \in \partial\Omega$ we can construct an exterior tangent ball

$$B_\rho \equiv \{w : \|w - \tilde{w}\| < \rho\} \quad \text{where} \quad \|w_0 - \tilde{w}\| = \rho$$

(i.e., ∂B_ρ is tangent to $\partial\Omega$ at w_0 and $B_\rho \cap \Omega = \phi$). Fix $w_0 \in \partial\Omega$ and the corresponding ball B_ρ. We can choose coordinates so that the center of B_ρ is $\tilde{w} = 0$, i.e, $B_\rho = \{w : \|w\| < \rho\}$ and $\|w_0\| = \rho$. Choose R sufficiently large so that $\Omega \subset B_R = \{w : \|w\| < R\}$. We let $m = \max_{w \in \bar{\Omega}} |\Delta_{(w)}\phi(w)|$ and define, for $w \in \bar{\Omega}$,

$$g_1(w) \equiv \phi(w) - \frac{m}{n}\left[(R^2 - \rho^2)\frac{1/\rho^{2n-2} - 1/\|w\|^{2n-2}}{1/\rho^{2n-2} - 1/R^{2n-2}} - \|w\|^2 + \rho^2\right]$$

$$\equiv \phi(w) - \frac{m}{n}\psi(w)$$

and

$$g_2(w) \equiv \phi(w) + \frac{m}{n}\psi(w).$$

Direct calculation shows that $\psi(w_0) = 0$ and $\psi(w) \geq 0$ for $w \in \bar{\Omega}$; also, since $\Delta_{(w)} \dfrac{1}{\|w\|^{2n-2}} = 0$ if $w \neq 0$,

$$\Delta_{(w)}\psi(w) = -\Delta_{(w)}(\|w\|^2) = -n.$$

Hence

$$\Delta_{(w)}g_1(w) = \Delta_{(w)}\phi(w) + \max_{w \in \bar{\Omega}} |\Delta_{(w)}\phi(w)| \geq 0 \quad \text{for} \quad w \in \bar{\Omega}$$

while

$$g_1(w) \leq \phi(w) \quad \text{for} \quad w \in \bar{\Omega}.$$

Similarly, $\Delta_{(w)}g_2(w) \leq 0$ for $w \in \bar{\Omega}$ while $g_2(w) \geq \phi(w)$ for $w \in \bar{\Omega}$. Since g_1 is subharmonic in $\bar{\Omega}$ and $g_1 \leq \phi$ on $\partial\Omega$, by the maximum principle, $g_1 - H_\phi^\Omega$ has a maximum at w_0; hence

$$\frac{\partial g_1}{\partial n_w}(w_0) \geq \frac{\partial H_\phi^\Omega}{\partial n_w}(w_0). \tag{5.7}$$

Similarly, the subharmonic function $H_\phi^\Omega - g_2$ has a maximum at w_0 and

$$\frac{\partial H_\phi^\Omega}{\partial n_w}(w_0) \geq \frac{\partial g_2}{\partial n_w}(w_0). \tag{5.8}$$

Since ∂B_ρ and $\partial\Omega$ are tangent at w_0, $\dfrac{\partial}{\partial n_w} = \dfrac{-\partial}{\partial r}$ at w_0. Thus

$$\frac{\partial \psi}{\partial n_w}(w_0) = -\left[\frac{(2n-2)(R^2-\rho^2)}{\rho[1-(\frac{\rho}{R})^{2n-2}]} - 2\rho\right] \equiv -\alpha < 0.$$

Equations (5.7) and (5.8) yield that

$$(5.9) \qquad \frac{\partial \phi}{\partial n_w}(w_0) + \frac{m}{n}\alpha \geq \frac{\partial H^{\Omega}_{\phi}}{\partial n_w}(w_0) \geq \frac{\partial \phi}{\partial n_w}(w_0) - \frac{m}{n}\alpha .$$

Setting $C = \frac{\alpha}{n}$ in (5.9) yields (1). Since $\phi = H^{\Omega}_{\phi}$ on $\partial\Omega$, all tangential derivatives of ϕ and H^{Ω}_{ϕ} agree on $\partial\Omega$ so that

$$\|\mathrm{Grad}_{(w)}(H^{\Omega}_{\phi} - \phi)(w_0)\| = \frac{1}{2}\left|\frac{\partial}{\partial n_w}(H^{\Omega}_{\phi} - \phi)(w_0)\right|$$

and hence by (5.9)

$$\|\mathrm{Grad}_{(w)}(H^{\Omega}_{\phi})(w_0)\| \leq \|\mathrm{Grad}_{(w)}\phi(w_0)\| + Cm,$$

which is (2). ∎

<u>Remarks</u>. 1. Since H^{Ω}_{ϕ} is harmonic, each function $\left|\frac{\partial H^{\Omega}_{\phi}}{\partial w_\alpha}\right|$ is subharmonic $(\alpha = 1,\cdots,n)$ so that

$$(5.10) \qquad \left|\frac{\partial H^{\Omega}_{\phi}}{\partial w_\alpha}(w)\right| \leq \max_{w_0 \in \partial\Omega} \|\mathrm{Grad}_{(w)}\phi(w_0)\| + Cm, \quad w \in \bar{\Omega} .$$

2. From the proof, it follows that if $\partial\Omega$ is not smooth, (5.6) still remains valid at any smooth boundary point w, where the constant C depends on ρ, the radius of an externally tangent ball to $\partial\Omega$ at w, $R = \rho$ + diameter of Ω, and n.

3. Clearly this result is valid in R^m, $m \geq 3$ (and even $m = 2$, with minor modifications); we have stated the C^n-version since it is the precise result needed for our applications.

In order to apply the lemma to $\phi(w) \equiv F(\xi,w)$ near $w \in \partial D(\xi)$, we need to estimate F, $\frac{\partial F}{\partial w_\alpha}$, and $\frac{\partial^2 F}{\partial w_\alpha \partial \bar{w}_\alpha}$ there. The precise estimates will be stated and proved in Step 3. We first need to refine the estimate from Step 3 of Chapter 4:

$$\|\mathrm{Grad}_{(w)} g(\xi,w)\| \le \frac{K}{\|w\|^{2n-1}}, \quad \xi \in D_1, \quad w \in \partial D(\xi), \tag{4.2}$$

where $D_1 \equiv \{\xi \in \mathbb{C}^n : \mathrm{dist}(\xi,\partial D) < 1\}$ and $K > 0$ is a constant independent of ξ and w. In order to do this, we need two rather technical but elementary results which we state without proof.

Let $x = (x_1,\cdots,x_m)$ denote coordinates in R^m and $x' = (x_1,\cdots,x_{m-1}) \in R^{m-1}$. Let $\{\phi_p \equiv \phi_p(x')\}$ be a sequence of C^∞ functions defined in $\{x' \in R^{m-1} : \|x'\| < 1\}$ which converge uniformly, together with all derivatives of order $\le N$ $(N \ge 6)$ to a function $\phi(x')$ of class C^N. Set

$\Omega_p \equiv \{x \in R^m : \|x\| < 1, \ x_m < \phi_p(x')\}$, $p = 1,2,\cdots$ and

$\Omega \equiv \{x \in R^m : \|x\| < 1, \ x_m < \phi(x')\}$.

Proposition 5.1. Let $\{u_p = u_p(x)\}$ be a sequence of functions such that u_p is harmonic in Ω_p and continuous up to $\partial\Omega_p \cap \{x : \|x\| < 1\}$ and such that u_p satisfies

(a) $0 \le u_p(x) \le 1$, $x \in \Omega_p$;

(b) $u_p(x) = 0$ if $x \in \partial\Omega_p \cap \{x : \|x\| < 1\}$,

(c) $\{u_p\}$ converges uniformly on compact subsets of Ω to a harmonic function u.

Then if $\{x^{(p)}\}$ is a sequence of points with

$$x^{(p)} \in \Omega_p \cup [\partial\Omega_p \cap \{x : \|x\| < 1\}]$$

which converges to a point

$$x^{(0)} \in \Omega \cup [\partial\Omega \cap \{x : \|x\| < 1\}],$$

we have

$$\lim_{p\to+\infty} u_p(x^{(p)}) = u(x^{(0)}); \quad \lim_{p\to+\infty} \frac{\partial u_p}{\partial x_i}(x^{(p)}) = \frac{\partial u}{\partial x_i}(x^{(0)}) ;$$

$$\lim_{p\to+\infty} \frac{\partial^2 u_p}{\partial x_i \partial x_j}(x^{(p)}) = \frac{\partial^2 u}{\partial x_i \partial x_j}(x^{(0)}); \quad \lim_{p\to+\infty} \frac{\partial^3 u}{\partial x_i \partial x_j \partial x_k}(x^{(p)}) = \frac{\partial^3 u}{\partial x_i \partial x_j \partial x_k}(x^{(0)})$$

for $i,j,k = 1,\cdots,n$.

Remark. Clearly the stronger the convergence of $\{\phi_p\}$, the more derivatives of $\{u_p\}$ which converge as $x^{(p)} \longrightarrow x^{(0)}$. Proposition 5.1 can be rigorously proved using Fredholm's theory of integral equations.

We will also need the following version of a tubular neighborhood theorem.

Proposition 5.2. Let $D \subset R^m$ be a bounded domain with smooth boundary and let $N \geq 6$ be a positive integer. There exist $0 < r < 1$ and $M > 1$ such that for any point $x_0 = (x_0', x_{0m})$ in the neighborhood

$$U = U_r \equiv \bigcup_{x\in\partial D} \{x + tn_x : -r < t < r\}$$

of ∂D (n_x = outer unit normal to ∂D at x),

$$\partial D \cap B(x_0, r) = \{x \in \partial D : \|x - x_0\| < r\}$$

can be represented, after a rotation and translation of coordinates so that

$n_{x^*} = (0,\cdots,0,1) \equiv e_m$, where $x_0 = x^* + tn_{x^*}$ and $x^* \in \partial D$ in the form $x_m = \phi(x')$ where

(a) $\phi(x')$ is C^∞ in $\{x' \in R^{m-1} : \|x'-x_0'\| < r\}$ with $\phi(x_0') = x_{0m}-t$

and

(b) all partial derivatives of ϕ of order $\le N$ are bounded in absolute value on $\{x' : \|x'-x_0'\| < r\}$ by M.

Remark. Given $\varepsilon > 0$, we can find r, M as above such that, in addition to (a) and (b), we also have

(b)′ $|\frac{\partial \phi}{\partial x_i}| < \varepsilon$, $i = 1,\cdots,m-1$, on $\{x' : \|x'-x_0'\| < r\}$.

This will be useful in Step 3.

The importance of Proposition 5.2 is that the numbers r and M do not depend on the point x_0. We now use the above propositions to improve the estimate (4.2). Let $w = (w_1,\cdots,w_n) = (y_1,y_2,\cdots,y_{2n})$.

Step 2. There exists a constant $K > 0$, independent of ξ and w, such that

$$(5.11)\quad \begin{cases} (1) & \left|\frac{\partial g}{\partial y_i}(\xi,w)\right| < \frac{K}{\|w\|^{2n-1}} \\ (2) & \left|\frac{\partial^2 g}{\partial y_i \partial y_j}(\xi,w)\right| < \frac{K}{\|w\|^{2n}} \\ (3) & \left|\frac{\partial^3 g}{\partial y_i \partial y_j \partial y_k}(\xi,w)\right| < \frac{K}{\|w\|^{2n+1}}, \quad i,j,k = 1,\cdots,2n, \end{cases}$$

for all $\xi \in D_1 \equiv \{\xi \in \mathbb{C}^n : \operatorname{dist}(\xi,\partial D)\} < 1$ and $w \in \overline{D(\xi)} - \{0\}$.

Remark. If $\xi \in \partial D$, the explicit formula

$$g(\xi, w) = \frac{1}{\|w\|^{2n-2}} - \frac{1}{\|w-\bar{N}_\xi\|^{2n-2}}$$

for the half-space $D(\xi)$ implies the existence of a constant $K > 0$, independent of $\xi \in \partial D$, such that (1) - (3) hold. Thus in the proof of Step 2 we may assume $\xi \in D \cup D'$.

Proof of Step 2. Fix a positive integer N and choose r, M and $U = U_r$ as in Proposition 5.2. Choose $0 < \delta < 1$ sufficiently small so that

$$\max \{\delta d(D), \frac{\delta}{1-\delta} (d(D) + 2)\} < r$$

where $d(D)$ = diameter of D. Note this implies that if $\xi, z_0 \in D_1$ and $\xi_0 \in \partial D$ satisfy

(5.12) $\|z_0 - \xi_0\| \leq \delta\|z_0 - \xi\|$, then $\delta\|z_0 - \xi\| < r$ so that $\|z_0 - \xi_0\| < r$.

Fix $\xi \in D_1 - \partial D$ and $w_0 \in D(\xi)$. Consider the ball

$$B(w_0) \equiv \{w \in \mathbb{C}^n : \|w - w_0\| < \delta\|w_0\|\}$$

Case 1. $B(w_0) \subset\subset D(\xi)$.

In this situation, since $0 \notin B(w_0)$ and $0 < g(\xi, w) < \frac{1}{\|w\|^{2n-2}}$ for all $w \in D(\xi)$, from the Poisson integral formula applied to $g(\xi, w)$ on $B(w_0)$, there exists $c_n > 0$ such that

$$(5.13)\quad \begin{cases} (1)\ \left|\dfrac{\partial g}{\partial y_i}(\xi,w_0)\right| < c_n\left[\max_{w\in\partial B(w_0)} g(\xi,w)\right]\dfrac{1}{\delta\|w_0\|} \le \dfrac{c_n}{(1-\delta)^{2n-2}\delta}\,\dfrac{1}{\|w_0\|^{2n-1}} \\ (2)\ \left|\dfrac{\partial^2 g}{\partial y_i\partial y_j}(\xi,w_0)\right| \le \dfrac{c_n}{(1-\delta)^{2n-2}\delta^2}\,\dfrac{1}{\|w_0\|^{2n}} \\ (3)\ \left|\dfrac{\partial^3 g}{\partial y_i\partial y_j\partial y_k}(\xi,w_0)\right| \le \dfrac{c_n}{(1-\delta)^{2n-2}\delta^3}\,\dfrac{1}{\|w_0\|^{2n+1}} \end{cases}$$

Since δ depends only on D and is independent of ξ and w_0, this proves Step 2 for Case 1.

Case 2. $B(w_0) \not\subset D(\xi)$.

The idea in this case is to transform to the setting of Proposition 5.1. Assume, by way of contradiction, that (5.11) does not hold; hence, we get a sequence of points (ξ_p, w_p) at which the derivatives of g get big; we then use the conclusion of Proposition 5.1 to derive a contradiction at a limit point of $\{(\xi_p, w_p)\}$.

We consider the transformation $w = T_\xi(z) = \dfrac{z-\xi}{-\psi(\xi)}$ and its inverse $z = T_\xi^{-1}(w) = \xi - \psi(\xi)w$. Set $z_0 = T_\xi^{-1}(w_0)$ and

$$b(z_0,\xi) = T_\xi^{-1}(B(w_0)) = \{z \in \mathbb{C}^n : \|z - z_0\| < \delta\|z_0 - \xi\|\}.$$

Since $B(w_0) \not\subset D(\xi)$, we can find a point $\xi_0 \in \partial D$ with

$$\|\xi_0 - z_0\| \le \delta\|z_0 - \xi\|.$$

By (5.12), $\|\xi_0 - z_0\| < r$ so that $z_0 \in U = U_r$. Hence we can find $\xi^x \in \partial D$ such that $z_0 = \xi^x + tn_{\xi^x}$ where

$$(5.14)\qquad 0 < |t| < \delta\|z_0 - \xi\| < r.$$

Let $\eta^x = T_\xi(\xi^x) \in \partial D(\xi)$. Via a rotation of the w-coordinates, we may assume $n_{\xi^x} = e_{2n} = (0,\cdots,0,1)$ so that the unit outer normal vector to

$\partial D(\xi)$ at η^x is also e_{2n} (if $\xi \in D$) or $-e_{2n}$(if $\xi \in D'$). Then

$$w_0 - \eta^x = (0,\cdots,0, \frac{t}{-\psi(\xi)})$$

and the tangent plane Π of $\partial D(\xi)$ at η^x is given by

$$\Pi = \{y = (y_1,\cdots,y_{2n}) : y_{2n} = \operatorname{Im} \eta_n^x\}$$

To transform the above setting to the unit ball, we let

$$W = S_{w_0}(w) \equiv \frac{w - w_0}{\delta\|w_0\|}$$

and

$$\Omega(\xi,w_0) \equiv S_{w_0}(D(\xi) \cap B(w_0)).$$

To apply Proposition 5.1, we first note that since $z_0 \in U = U_r$, $\partial D \cap \{z : \|z - z_0\| < r\}$ can be represented in the form

$$x_{2n} = \phi(x')$$

as in Proposition 5.2, where $\phi(x')$ is C^∞ if $\|x'-x_0'\| < r$ and all partial derivatives of ϕ of order $\le N$ are bounded in absolute value by M. Here, $z_k = x_{2k-1} + ix_{2k}$, $k = 1,\ldots,n$. Since

$$(S_{w_0} \circ T_\xi)^{-1}(W) = z = -\psi(\xi)w_0 + \xi - \psi(\xi)\,\delta\|w_0\|W,$$

the surface

$$\{x = (x',x_{2n}) : x_{2n} = \phi(x'),\quad \|x'-x_0'\| < r\}$$

is mapped by $S_{w_0} \circ T_\xi$ onto the surface

$$C \equiv \{Y \equiv (Y',Y_{2n}) : Y_{2n} = \Phi(Y'),\quad \|Y'\| < R\}$$

where, letting $w_0 = (y_0',(y_0)_{2n})$, $\xi = (\xi',\xi_{2n})$,

$$\Phi(Y') = \frac{\phi(-\psi(\xi)y_0' + \xi' - \psi(\xi)\delta\|w_0\|Y')}{-\psi(\xi)\delta\|w_0\|} + \frac{\psi(\xi)(y_0)_{2n} - \xi_{2n}}{-\psi(\xi)\delta\|w_0\|}$$

and

$$R = R(\xi, w_0) = \frac{r}{\psi(\xi)\delta\|w_0\|} = \frac{r}{\delta\|z_0-\xi\|} > 1$$

by (5.14). In particular,

$$\partial\Omega(\xi, w_0) \cap \{W : \|W\| < 1\} \subset C.$$

By using the properties of ϕ and the explicit formula for Φ above, it follows that

$$\partial\Omega(\xi, w_0) \cap \{W : \|W\| < 1\}$$

is represented in the form $Y_{2n} = \Phi(Y')$ where Φ is of class C^∞ in $\{Y': \|Y'\| < 1\}$ and satisfies

(a) $0 < \Phi(0) < 1$ by (5.14) and

(b) $\left|\frac{\partial^\alpha \Phi}{\partial Y^\alpha}\right| < M$ for all $\alpha = (\alpha_1, \cdots, \alpha_{2n})$ with $|\alpha| \le N$ if $\|Y'\| < 1$.

Therefore, as long as $\xi \in D_1 - \partial D$ and $w_0 \in D(\xi)$ are such that $B(w_0) \not\subset D(\xi)$, i.e., Case 2 applies, then the constant M is independent of ξ, w_0; in addition, the domain of definition of Φ can always be taken to contain $\{Y': \|Y'\| < 1\}$ for each pair (ξ, w_0).

We now complete the proof of Step 2 by contradiction. Since the proofs of (1)-(3) in (5.11) are similar, we will only prove (2). Thus assume (2) is not true. Then there exist sequences

$$\{\xi_p\} \subset D_1 - \partial D \quad\text{and}\quad \{w_p\}, \quad w_p \subset \overline{D(\xi_p)} - \{0\},$$

with

$$\lim_{p\to+\infty} \left|\frac{\partial^2 g}{\partial y_i \partial y_j}(\xi_p, w_p)\right| \|w_p\|^{2n} = +\infty . \tag{5.15}$$

Since each function $g(\xi_p, w_p)$ is of class C^4 for w up to $\partial D(\xi_p)$

(Chapter 4, Step 1 (1)), we may assume in (5.15) that $w_p \in D(\xi_p) - \{0\}$. Fix p. Then either Case 1 or Case 2 occurs for the point (ξ_p, w_p), i.e., either $B(w_p) \equiv \{w \in \mathbb{C}^n : \|w - w_p\| < \delta\|w_p\|\} \subset\subset D(\xi_p)$ or not. If Case 1 occurs, by (5.13) we have

$$\left|\frac{\partial^2 g}{\partial y_i \partial y_j} (\xi_p, w_p)\right| \leq \frac{c_n}{(1-\delta)^{2n-2}\delta^2} \frac{1}{\|w_p\|^{2n}} .$$

Since δ is independent of p, (5.15) implies that Case 1 occurs for only finitely many (ξ_p, w_p). Thus, for sufficiently large p, Case 2 occurs for all (ξ_p, w_p) and hence either D or D' (or both) contain infinitely many points ξ_p. For simplicity, we assume $\xi_p \in D_1 \cap D$ and Case 2 occurs for all (ξ_p, w_p). Then for each (ξ_p, w_p), after a rotation of coordinates and application of the transformation $S_{w_p} \circ T_{\xi_p}$, we obtain a domain

$$\Omega_p = \Omega(\xi_p, w_p) \equiv \{Y = (Y', Y_{2n}) : \|Y'\| < 1,\ Y_{2n} < \Phi_p(Y')\} \cap \{\|W\| < 1\}$$

where Φ_p is of class C^∞ in $\{Y' : \|Y'\| \leq 1\}$ and satisfies

(a) $0 < \Phi_p(0) < 1$ and

(b) $\left|\frac{\partial^\alpha \Phi_p}{\partial Y^\alpha}\right| < M$ for all $|\alpha| \leq N$, if $\|Y'\| < 1$.

Since M is independent of p, by the Arzela-Ascoli theorem we can find a subsequence, which we again call $\{\Phi_p\}$, and a function Φ of class C^N on $\{Y' : \|Y'\| < 1\}$ such that $\{\Phi_p\}$ converges uniformly to Φ on compact subsets of $\{Y' : \|Y'\| < 1\}$, together with all partial derivatives of order $\leq N$.

We now set

$$\Omega = \{Y = (Y', Y_{2n}) : \|Y'\| < 1,\ Y_{2n} < \Phi(Y')\} \cap \{\|W\| < 1\}$$

Note since $0 \in \Omega_p$, $p = 1, 2, \cdots$, we have $0 \in \bar{\Omega}$. We want to apply

Proposition 5.1 to the functions

$$u_p = u_p(W) \equiv \|w_p\|^{2n-2}(1 - \delta)^{2n-2}\, g(\xi_p, w) \tag{5.16}$$

for $W = \dfrac{w - w_p}{\delta\|w_p\|} \in \Omega_p$. Since

$$0 < g(\xi_p, w) < \frac{1}{\|w\|^{2n-2}}, \quad w \in D(\xi_p),$$

if $w \in B(w_p) \equiv \{w : \|w - w_p\| < \delta\|w_p\|\}$,

$$0 < g(\xi_p, w) < \frac{1}{(1-\delta)^{2n-2}\,\|w_p\|^{2n-2}}$$

so that u_p is harmonic in Ω_p, continuous up to $\partial\Omega_p$, $0 \le u_p(W) \le 1$ in Ω_p, and $u_p(W) = 0$ on $\partial\Omega_p \cap \{\|W\| < 1\}$. By Harnack's Theorem, a subsequence $\{u_p\}$ converges uniformly on compact subsets of Ω to a harmonic function $u(W)$ on Ω. From Proposition 5.1, it follows that

$$\lim_{p\to+\infty} \frac{\partial^2 u_p}{\partial Y_i \partial Y_j}(0) = \frac{\partial^2 u}{\partial Y_i \partial Y_j}(0)$$

which is finite. Hence from the definition of u_p in (5.16)

$$\overline{\lim_{p\to+\infty}} \left|\frac{\partial^2 g}{\partial y_i \partial y_j}(\xi_p, w_p)\right| \|w_p\|^{2n} \le \frac{\delta^2}{(1-\delta)^{2n-2}} \left|\frac{\partial^2 u}{\partial Y_i \partial Y_j}(0)\right| < +\infty,$$

contradicting (5.15). Thus Step 2 is proved. ∎

In order to obtain our required estimates on F, $\dfrac{\partial F}{\partial w_\alpha}$, and $\dfrac{\partial^2 F}{\partial w_\alpha \partial \bar{w}_\alpha}$, which is the content of our next step, we make some comments which will be needed.

Let $\xi_0 \in \partial D$ and consider a sequence of points $\{\xi_p\}$ in D (or D′) with

$$\lim_{p\to+\infty} \xi_p = \xi_0 .$$

We have often used the transformations

$$T_{\xi_p}(z) \equiv \frac{z - \xi_p}{-\psi(\xi_p)}$$

which transform D to $T_{\xi_p}(D) \equiv D(\xi_p)$; hence we obtain a sequence of domains $D(\xi_p)$, similar to D, which tend to the half-space $D(\xi_0)$ as $p \to +\infty$. Instead we now take a sequence of positive numbers $\{k_p\}$ with $\lim_{p\to+\infty} k_p = 0$ such that for each p, the ball

$$b_p \equiv \{z : \|z - \xi_p\| < k_p\}$$

intersects ∂D, i.e., $b_p \cap \partial D \neq \phi$. We assume, via translation and rotation, that $n_{\xi_p^x} = (0,\dots,1) = e_{2n}$ where $\xi_p^x = \xi_p + tn_{\xi_p^x} \in \partial D$. Again, since $k_p \to 0$, the transformations

$$S_p(z) \equiv W = \frac{z - \xi_p}{k_p}$$

transform D to $D_p \equiv S_p(D)$, and we obtain a sequence of domains D_p, similar to D, which tend to the half-space $D(\xi_0)$. In this case, using a similar procedure to that used in Step 2, we can prove the following.

<u>Proposition</u> <u>5.3</u>. Given ξ_0, $\{\xi_p\}$, $\{k_p\}$ as above, and a positive integer N, there exists a subsequence of $\{\xi_p\}$, a sequence of positive numbers $\{r_p\}$ with $\lim_{p\to+\infty} r_p = +\infty$, and a sequence of functions $\{\Phi_p\}$ with $\Phi_p = \Phi_p(Y')$ of class C^∞ in $\{Y' : \|Y'\| < r_p\}$ such that $\partial D_p \cap \{\|W\| < r_p\}$ is written in the form

$$Y_{2n} = \Phi_p(Y'), \text{ and}$$

(1) $\Phi_p(Y')$ converges uniformly on compact sets to a constant, c.

(2) All partial derivatives of Φ_p of order $\le N$ converge uniformly on compact subsets to 0.

<u>Remark</u>. (1) implies that the domains $D_p \cap \{\|W\| < r_p\}$ tend to the half-space $\tilde{D} \equiv \{W \in \mathbb{C}^n : Y_{2n} < c\}$.

<u>Proof</u>. Fix $\varepsilon > 0$. By the remark after Proposition 5.2, we can find $r_\varepsilon, M_\varepsilon$, and $U = U_{r_\varepsilon}$ such that for points $\xi \in U$,

$$\partial D \cap B(\xi, r_\varepsilon) = \{\tilde{\xi} \in \partial D : \|\xi - \tilde{\xi}\| < r_\varepsilon\}$$

can be represented in local coordinates as $x_n = \phi(x')$ with

(a) $\phi(x')$ is C^∞ in $\{x' : \|x'-x_0'\| < r_\varepsilon\}$, $\phi(x_0') = (x_0)_{2n} + t$, and

(5.17)

(b) $\left|\frac{\partial\phi}{\partial x_i}(x')\right| < \varepsilon$; $\left|\frac{\partial^\alpha \phi}{\partial x^\alpha}(x')\right| < M_\varepsilon$, $1 < |\alpha| \le N$ if $\|x'-x_0'\| < r_\varepsilon$.

Since $\xi_p \longrightarrow \xi_0 \in \partial D$, $\xi_p \in U$ for sufficiently large p. Thus we may assume $\xi_p \in U \cap D$ for all p and (5.17) holds for a sequence of functions $\{\phi_p\}$ where $\phi_p = \phi_p(\xi_p')$ is defined for $\|\xi_p' - \xi_0'\| < r_\varepsilon$.

Define

$$\Phi_p(Y') \equiv \frac{\phi_p(\xi_p' + k_p Y')}{k_p} - \frac{(\xi_p)_{2n}}{k_p} \tag{5.18}$$

for

$$\left\{Y' : \|Y'\| < r_p \equiv \frac{r_\varepsilon}{k_p}\right\}. \tag{5.19}$$

Then $\lim_{p\to+\infty} k_p = 0$ implies that $\lim_{p\to+\infty} r_p = +\infty$. Note that

$|\Phi_p(0)| = \frac{|\phi_p(\xi_p')-(\xi_p)_{2n}|}{k_p} = \frac{|t|}{k_p} < 1$ since $b_p \cap \partial D \neq \phi$. From (5.17) and

(5.18) it follows that

$$\left|\frac{\partial\Phi_p}{\partial Y_i}\right| < \varepsilon; \quad \left|\frac{\partial^\alpha \Phi_p}{\partial Y^\alpha}\right| < M_\varepsilon k_p^{|\alpha|-1}, \quad 1 < |\alpha| \le N \quad \text{if} \quad \|Y'\| < r_p.$$

Thus by Arzela-Ascoli we can find a subsequence of p's such that $\{\Phi_p\}$, $\left\{\frac{\partial\Phi_p}{\partial Y_i}\right\}$ converge uniformly to Φ, $\frac{\partial\Phi}{\partial Y_i}$ on compact sets while $\frac{\partial^\alpha\Phi_p}{\partial Y^\alpha}$ converges uniformly to 0 on compact sets if $2 \le |\alpha| \le N$. Thus to prove (1) and (2) it suffices to show that $\frac{\partial\Phi}{\partial Y_i} \equiv 0$, i.e., $\Phi \equiv$ constant. To do this, note that given any $\varepsilon > 0$, we can find r_ε, M_ε such that for p sufficiently large

$$(5.20) \qquad \xi_p \in U = U_{r_\varepsilon} \quad \text{and} \quad \left|\frac{\partial\Phi_p}{\partial Y_i}(Y')\right| < \varepsilon \quad \text{if} \quad \|Y'\| < \frac{r_\varepsilon}{k_p}$$

Since $\lim_{p\to+\infty} \frac{r_\varepsilon}{k_p} = +\infty$, (5.20) implies that $\frac{\partial\Phi}{\partial Y_i}(Y') \equiv 0$ for all Y' and the proposition is proved. ∎

To get the desired estimates on F, $\frac{\partial F}{\partial w_\alpha}$, and $\frac{\partial^2 F}{\partial w_\alpha \partial \bar{w}_\alpha}$ near $\partial D(\xi)$, we first describe the set of points (ξ, w) at which we want these estimates. From the definition of F,

$$(4.19) \qquad F(\xi,w) \equiv \frac{-\frac{\partial f}{\partial \xi_\nu}(\xi,w)}{\|\mathrm{Grad}_{(w)} f(\xi,w)\|} \|\mathrm{Grad}_{(w)} g(\xi,w)\|,$$

it follows that F is defined for all $\xi \in \mathbb{C}^n$ and $w \in \overline{D(\xi)}$ except at those points $(\xi,w) \in \mathbb{C}^n \times \{0\}$ (the pole of $g(\xi,w)$) and points where $\mathrm{Grad}_{(w)} f(\xi,w) = 0$. Given $r > 0$, $\xi \in \mathbb{C}^n$, and $w_0 \in \partial D(\xi)$, we define

$$B_r(w_0) \equiv \{w \in \mathbb{C}^n : \|w - w_0\| < r\|w_0\|\} ;$$

$$E_r(\xi, w_0) \equiv B_r(w_0) \cap D(\xi) ;$$

and

$$E_r(\xi) \equiv \bigcup_{w_0 \in \partial D(\xi)} E_r(\xi, w_0).$$

Note that $E_r(\xi)$ is a 'collar' about $\partial D(\xi)$ (lying inside $D(\xi)$) which does not contain 0 if $r < 1$. Finally, given any set $U \subset \mathbb{C}^n$, define

$$\mathcal{E}_r(U) \equiv \bigcup_{\xi \in U} (\xi,\ E_r(\xi)).$$

Step 3. There exist $0 < r < 1$, an open neighborhood U of ∂D with $U \subset\subset D_1$, and $M > 0$ such that

$$\text{(5.21)} \qquad \begin{cases} (1) & |F(\xi, w)| < \dfrac{M}{\|w\|^{2n-3}} \\[2ex] (2) & \left|\dfrac{\partial F}{\partial y_i}(\xi, w)\right| < \dfrac{M}{\|w\|^{2n-2}} \\[2ex] (3) & \left|\dfrac{\partial^2 F}{\partial y_i \partial y_j}(\xi, w)\right| < \dfrac{M}{\|w\|^{2n-1}} \end{cases}$$

for all $(\xi, w) \in \mathcal{E}_r(U)$, $i, j = 1, \cdots, 2n$ (recall $w_\alpha = y_{2\alpha-1} + iy_{2\alpha}$).

Remark. For $\xi \in \partial D$, the explicit formula for $g(\xi, w)$ and hence for $F(\xi, w)$ yields the existence of a constant M, independent of $\xi \in \partial D$, such that

$$(1)' \qquad |F(\xi, w)| < \frac{M}{\|w\|^{2n-2}}$$

$$(2)' \qquad \left|\frac{\partial F}{\partial y_i}(\xi, w)\right| < \frac{M}{\|w\|^{2n-1}}$$

$$(3)' \qquad \left|\frac{\partial^2 F}{\partial y_i \partial y_j}(\xi, w)\right| < \frac{M}{\|w\|^{2n}}$$

for all $w \in \overline{D(\xi)}$ with $\|w\| > r_0$, where $r_0 > 0$ with

$$\{w : \|w\| < r_0\} \subset\subset D(\xi)$$

for all $\xi \in \partial D$ (even for all $\xi \in \mathbb{C}^n$). Such r_0 exists since $0 \in D(\xi)$ and $f(\xi, w)$ is smooth. Thus, as in the proof of Step 2, we may assume that $\xi \in \mathbb{C}^n - \partial D$ in the proof of Step 3.

Proof of Step 3. We first look at the correspondence between the sets $B_r(w_0)$, $E_r(\xi, w_0)$, and $E_r(\xi)$ in the w-variables and sets in the original z-variables via the transformation $w = T_\xi(z) = \frac{z - \xi}{-\psi(\xi)}$. Fix $0 < r < 1$, $\xi \in D \cup D'$, and $z_0 \in \partial D$. Define

$$b_r(\xi, z_0) \equiv \{z \in \mathbb{C}^n : \|z - z_0\| < r\|z_0 - \xi\|\} ;$$

$$\beta_r(\xi, z_0) \equiv \begin{cases} b_r(\xi, z_0) \cap D & \text{if } \xi \in D \\ b_r(\xi, z_0) \cap D' & \text{if } \xi \in D' \end{cases} ; \text{ and}$$

$$\beta_r(\xi) = \bigcup_{z_0 \in \partial D} \beta_r(\xi, z_0).$$

Note that since $r < 1$, $\beta_r(\xi)$ is a 'collar' about ∂D (lying in D if $\xi \in D$ and lying in D' if $\xi \in D'$) which does not contain the point ξ. Given any set $U \subset \mathbb{C}^n$, we define $\beta_r(U - \partial D) \equiv \bigcup_{\xi \in U - \partial D} (\xi, \beta_r(\xi))$. Under the transformation $w_0 = T_\xi(z_0)$, the sets $b_r(\xi, z_0)$, $\beta_r(\xi, z_0)$ and $\beta_r(\xi)$ correspond to the sets $B_r(w_0)$, $E_r(\xi, w_0)$ and $E_r(\xi)$. Thus, under the mapping $T(\xi, z) \equiv (\xi, w) = (\xi, T_\xi(z))$, the sets $\beta_r(U - \partial D)$ and $\mathcal{E}_r(U - \partial D)$ correspond where $U \subset \mathbb{C}^n$.

To get a set U with $\partial D \subset\subset U \subset\subset D_1$ so that the estimates (5.21) hold, we first note that since $\operatorname{Grad} \psi(z) \neq 0$ for $z \in \partial D$, we can find an open neighborhood U_1 of ∂D and a constant $m > 0$ such that

(5.22) $$\|\text{Grad } \psi(z)\| \geq m \quad \text{for all} \quad z \in U_1.$$

Since D is bounded, we can find $0 < r_1 < 1$ and a neighborhood U_2 of ∂D such that

(5.23) $$\beta_{r_1}(U_2 - \partial D) \subset U_2 \times U_1.$$

Recalling that

$$\text{Grad}_{(w)} f(\xi, w) = \text{Grad}_{(z)} \psi(z)$$

(where $z = \xi - \psi(\xi)w$), it follows from (5.22) and (5.23) that

(5.24) $\|\text{Grad}_{(w)} f(\xi, w)\| \geq m$ for $(\xi, w) \in \mathcal{E}_{r_1}(U_2 - \partial D)$, hence in $\mathcal{E}_{r_1}(U_2)$

by continuity. We now show that with $U = U_2$, $r = r_1$, we can find $M > 0$ so that (5.21) (1) and (2) hold. This will be accomplished with the estimates on f and its derivatives from Chapter 4, Step 4, which involved a constant $A > 0$; and the estimates on g and its derivatives from Chapter 5, Step 2, which involved a constant $K > 0$. Applying these estimates for $(\xi, w) \in \mathcal{E}_{r_1}(U_2)$ to (4.19) we obtain

$$|F(\xi, w)| = \frac{\left|\frac{\partial f}{\partial \xi_\nu}(\xi, w)\right|}{\|\text{Grad}_{(w)} f(\xi, w)\|} \|\text{Grad}_{(w)} g(\xi, w)\| \leq \frac{A\|w\|^2}{m} \frac{K}{\|w\|^{2n-1}} \equiv \frac{M}{\|w\|^{2n-3}}$$

which proves (1). To prove (2), we compute, for each fixed $\xi \in \mathbb{C}^n$, the derivative $\frac{\partial F}{\partial y_i}$ at those points $w \in \mathbb{C}^n$ where this derivative is defined. Thus

$$(5.25) \quad \frac{\partial F}{\partial y_i} = \frac{\frac{-\partial^2 f}{\partial \xi_\nu \partial y_i}}{\|\mathrm{Grad}_{(w)} f\|} \|\mathrm{Grad}_{(w)} g\| + \frac{1}{4} \frac{\partial f}{\partial \xi_\nu} \frac{\sum_{k=1}^{2n} \frac{\partial f}{\partial y_k} \frac{\partial^2 f}{\partial y_k \partial y_i}}{\|\mathrm{Grad}_{(w)} f\|^3} \|\mathrm{Grad}_{(w)} g\|$$

$$- \frac{1}{4} \frac{\partial f}{\partial \xi_\nu} \frac{1}{\|\mathrm{Grad}_{(w)} f\|} \frac{\sum_{k=1}^{2n} \frac{\partial g}{\partial y_k} \frac{\partial^2 g}{\partial y_k \partial y_i}}{\|\mathrm{Grad}_{(w)} g\|}$$

provided w, $\mathrm{Grad}_{(w)} f(\xi, w)$, $\mathrm{Grad}_{(w)} g(\xi, w) \neq 0$. Using the above-mentioned estimates plus the fact that

$$\left|\frac{\partial g}{\partial y_k}\right| \Big/ \|\mathrm{Grad}_{(w)} g\| \leq 2,$$

we obtain, for $(\xi, w) \in \mathcal{E}_{r_1}(U_2)$,

$$\left|\frac{\partial F}{\partial y_i}\right| \leq \frac{A\|w\|}{m} \frac{K}{\|w\|^{2n-1}} + \frac{1}{4} A\|w\|^2 \frac{2n\ A \frac{A}{\|w\|}}{m^3} \frac{K}{\|w\|^{2n-1}}$$

$$+ \frac{1}{4} A\|w\|^2 \frac{1}{m} 2n \cdot 2 \frac{K}{\|w\|^{2n}} \equiv \frac{M_2}{\|w\|^{2n-2}}$$

which proves (2).

In order to prove (3), we differentiate (5.25) with respect to y_j and estimate as before. All terms except those of the form

$$\frac{\frac{\partial f}{\partial \xi_\nu}}{\|\mathrm{Grad}_{(w)} f\|} \frac{\frac{\partial^2 g}{\partial y_k \partial y_i} \frac{\partial^2 g}{\partial y_\ell \partial y_i}}{\|\mathrm{Grad}_{(w)} g\|} \quad \text{or} \quad \frac{\frac{\partial f}{\partial \xi_\nu}}{\|\mathrm{Grad}_{(w)} f\|} \frac{\frac{\partial g}{\partial y_k} \frac{\partial g}{\partial y_\ell} \frac{\partial^2 g}{\partial y_k \partial y_i} \frac{\partial^2 g}{\partial y_\ell \partial y_j}}{\|\mathrm{Grad}_{(w)} g\|^3}$$

can be estimated from above by $\frac{\text{const.}}{\|w\|^{2n-1}}$ in $\mathcal{E}_{r_1}(U_2)$. To finish the proof of (3) it suffices to prove the existence of constants $M_3 > 0$, $r_2 < r_1$ $(r_2 > 0)$, and an open neighborhood U_3 of ∂D with $U_3 \subset U_2$ such that

(5.26)
$$\frac{\left|\frac{\partial^2 g}{\partial y_i \partial y_j}(\xi,w)\right|}{\|\mathrm{Grad}_{(w)} g(\xi,w)\|} \le \frac{M_3}{\|w\|} \quad \text{in} \quad \mathcal{E}_{r_2}(U_3).$$

Since $g(\xi,w) = \psi(\xi)^{2n-2}\, G(\xi,z)$, where $z = \xi - \psi(\xi)w$, (5.26) is equivalent to

(5.27)
$$\frac{\left|\frac{\partial^2 G}{\partial x_i \partial x_j}(\xi,z)\right|}{\|\mathrm{Grad}_{(z)} G(\xi,z)\|} \le \frac{M_3}{\|z-\xi\|} \quad \text{in} \quad \mathcal{B}_{r_2}(U_3 - \partial D).$$

Note that (5.26) holds for $\xi \in \partial D$ from the explicit formula for $g(\xi,w)$ in this case.

We prove (5.27) by contradiction. If there exist no positive constants M_3, r_2 and open neighborhood U_3 of ∂D such that (5.27) holds, then we can choose a sequence of points $\{\xi_p\}$ in $D \cup D'$ with $\xi_p \to \partial D$ as $p \to +\infty$; and sequences $z_p \in \mathcal{B}_{1/p}(\xi_p)$, $z_{0p} \in \partial D$ satisfying

(5.28)
$$\|z_p - z_{0p}\| < \frac{1}{p}\|z_{0p} - \xi_p\|, \qquad p = 1,2,\cdots$$

such that

(5.29)
$$\frac{\left|\frac{\partial^2 G}{\partial x_i \partial x_j}(\xi_p,z_p)\right|}{\|\mathrm{Grad}_{(z)} G(\xi_p,z_p)\|} \ge \frac{p}{\|z_p-\xi_p\|}, \qquad p = 1,2,\cdots .$$

By taking subsequences, we may assume that

$$\lim_{p\to+\infty} \xi_p = \xi_0 \in \partial D \quad \text{and} \quad \lim_{p\to+\infty} z_{0p} = z_0 \in \partial D.$$

From (5.28) it follows that

$$\lim_{p\to+\infty} z_p = z_0 .$$

For simplicity, we may assume that all points ξ_p lie in D. We show first that $\xi_0 = z_0$. To prove this, we assume $\xi_0 \neq z_0$ and fix a point $a \in D$. For each $p = 1, 2, \cdots$, the function

$$h_p(z) \equiv \frac{G(\xi_p, z)}{G(\xi_p, a)}$$

is harmonic and positive in $D - \{\xi_p\}$; furthermore,

(5.30) $$h_p(a) = 1 \quad \text{and} \quad h_p(z) = 0 \quad \text{for} \quad z \in \partial D.$$

Since $\lim_{p \to +\infty} \xi_p = \xi_0 \in \partial D$, it follows from (5.30) and Harnack's principle that we can find a subsequence $\{h_p\}$ which converges uniformly on compact subsets of D to a positive harmonic function $h(z)$ which satisfies

$$h(a) = 1 \quad \text{and} \quad h(z) = 0 \quad \text{for} \quad z \in \partial D - \{\xi_0\}.$$

Since $z_0 \neq \xi_0$ and ∂D is smooth, we can take a small neighborhood U_{ξ_0} of ξ_0 with $z_0 \notin U_{\xi_0}$ and apply Proposition 5.1 to the functions h_p on $\Omega = \Omega_p = D - \bar{U}_{\xi_0}$. Note that $\exists\, M$ such that for p sufficiently large, $\sup \{h_p(z) : z \in \Omega\} < M$. We conclude that

(5.31) $$\begin{cases} \lim_{p \to +\infty} \|\mathrm{Grad}_{(z)} h_p(z_p)\| = \|\mathrm{Grad}_{(z)} h(z_0)\| \\ \lim_{p \to +\infty} \left| \dfrac{\partial^2 h_p}{\partial x_i \partial x_j}(z_p) \right| = \left| \dfrac{\partial^2 h}{\partial x_i \partial x_j}(z_0) \right| < +\infty . \end{cases}$$

By the Hopf lemma,

$$\|\mathrm{Grad}_{(z)} h(z_0)\| > 0.$$

Hence (5.31) implies that

$$\lim_{p\to+\infty} \frac{\left|\frac{\partial^2 G}{\partial x_i \partial x_j}(\xi_p, z_p)\right|}{\|\mathrm{Grad}_{(z)} G(\xi_p, z_p)\|} \|\xi_p - z_p\| = \lim_{p\to+\infty} \frac{\left|\frac{\partial^2 h_p}{\partial x_i \partial x_j}(z_p)\right|}{\|\mathrm{Grad}_{(z)} h_p(z_p)\|} \|\xi_p - z_p\|$$

$$= \frac{\left|\frac{\partial^2 h}{\partial x_i \partial x_j}(z_0)\right|}{\|\mathrm{Grad}_{(z)} h(z_0)\|} \|\xi_0 - z_0\| < +\infty,$$

which contradicts (5.29). Hence $\xi_0 = z_0$.

We now set

$$k_p \equiv \|\xi_p - z_{0p}\| > 0 \quad \text{and} \quad b_p \equiv \{z : \|z - \xi_p\| < k_p\}$$

Then $\xi_0 = z_0$ implies that

$$\lim_{p\to+\infty} k_p = 0;$$

furthermore, (5.28) implies that

$$b_p \cap \partial D \neq \emptyset.$$

From Proposition 5.3, we can write $\partial D_p \cap \{\|W\| < r_p\}$ in the form

$$Y_{2n} = \Phi_p(Y'), \quad \text{for} \quad \|Y'\| < r_p ,$$

where $S_p(z) \equiv W = \dfrac{z - \xi_p}{k_p} = (Y', Y_{2n})$, $D_p \equiv S_p(D)$, and $r_p \to +\infty$; $\{\Phi_p\}$ converges uniformly on compact sets to a constant c; and all partial derivatives of Φ_p of order $\leq N$ converge uniformly on compact subsets to 0. Let

$$\tilde{D} = \{W \in \mathbb{C}^n : Y_{2n} < c\}.$$

Then the domains $D_p \cap \{\|W\| < r_p\}$ tend to $\tilde{D}$. Setting

$$W_p = S_p(z_p) \quad \text{and} \quad W_{0p} = S_p(z_{0p}) ,$$

we see that $0 \in D_p$, $W_{0p} \in \partial D_p \cap \{\|W\| = 1\}$, and

$$\|W_p - W_{0p}\| = \frac{\|z_p - z_{0p}\|}{\|\xi_p - z_{0p}\|} < \frac{1}{p}$$

by (5.28). Thus, by choosing a subsequence if necessary, we have $0 \in \tilde{D} \cup \partial\tilde{D}$, $\lim_{p\to+\infty} W_{0p} = W_0$ for some $W_0 \in \partial\tilde{D} \cap \{\|W\| = 1\}$, and

$$\lim_{p\to+\infty} W_p = W_0 . \tag{5.32}$$

We will derive our contradiction by showing that $0 \notin \tilde{D} \cup \partial\tilde{D}$. This will be done in two steps.

<u>Step (i)</u>. $0 \notin \tilde{D}$.

<u>Proof of Step (i)</u>. If $0 \in \tilde{D}$, then

$$\tilde{g}_p(W) \equiv G(\xi_p, z)\, k_p^{2n-2}$$

is the Green function for $(D_p, 0)$. Again by Harnack's principle the functions $\{\tilde{g}_p\}$ converge uniformly on compact subsets of $\tilde{D} - \{0\}$ to the Green function $\tilde{g}(W)$ for $(\tilde{D}, 0)$. Thus, if we restrict W to lie in the ball $\{W : \|W - W_0\| < \frac{1}{2}\}$, then, analogous to (5.31), Proposition 5.1 and (5.32) imply that

$$\lim_{p\to+\infty} \|\mathrm{Grad}_{(W)}\tilde{g}_p(W_p)\| = \|\mathrm{Grad}_{(W)}\tilde{g}(W_0)\| > 0$$

and

$$\lim_{p\to+\infty} \frac{\partial^2\tilde{g}_p}{\partial Y_k \partial Y_\ell}(W_p) = \frac{\partial^2\tilde{g}}{\partial Y_k \partial Y_\ell}(W_0) \neq \infty .$$

Since $1 - \frac{1}{p} \leq \frac{\|z_p - \xi_p\|}{\|\xi_p - z_{0p}\|} \leq 1 + \frac{1}{p}$ by (5.28),

$$\overline{\lim_{p\to+\infty}} \frac{\left|\frac{\partial^2 G}{\partial x_k \partial x_\ell}(\xi_p, z_p)\right|}{\|\mathrm{Grad}_{(z)} G(\xi_p, z_p)\|} \|z_p - \xi_p\|$$

$$= \overline{\lim_{p\to+\infty}} \frac{\left|\frac{\partial^2 \tilde{g}_p}{\partial Y_k \partial Y_\ell}(W_p)\right|}{\|\mathrm{Grad}_{(W)} \tilde{g}_p(W_p)\|} \frac{\|z_p - \xi_p\|}{k_p} \le \frac{\left|\frac{\partial^2 \tilde{g}}{\partial Y_k \partial Y_\ell}(W_0)\right|}{\|\mathrm{Grad}_{(W)} \tilde{g}(W_0)\|} < +\infty,$$

which again contradicts (5.29). Thus $0 \notin \tilde{D}$.

<u>Step (ii)</u>. $0 \notin \partial\tilde{D}$.

<u>Proof of Step (ii)</u>. If $0 \in \partial\tilde{D}$, then $c = 0$, i.e.,

$$\tilde{D} = \{W \in \mathbb{C}^n : Y_{2n} < 0\}.$$

In this case, we fix a point

$$a \in \tilde{D} \cap \{W : \|W - W_0\| < \tfrac{1}{2}\}$$

and set

$$h_p(W) \equiv \tilde{g}_p(W)/\tilde{g}_p(a) .$$

Then h_p is positive and harmonic in $D_p - \{0\}$; $h_p = 0$ on ∂D_p and $h_p(a) = 1$. By choosing a subsequence if necessary, it follows from Harnack's principle that $\{h_p\}$ converges uniformly on compact subsets of $\tilde{D}$ to a positive harmonic function h; $h = 0$ on $\partial\tilde{D} - \{0\}$ and $h(a) = 1$. Furthermore,

$$h(W) = O\left(\frac{1}{\|W\|^{2n-2}}\right) \quad \text{as} \quad \|W\| \to +\infty .$$

For if we let $M_1 > \max\{h(W) : \|W\| = 1,\ Y_{2n} > 0\} > 0$, then for p sufficiently large,

$$2M_1 > \sup\{h_p(W) : \|W\| = 1,\ W \in D_p\} > 0.$$

Thus

$$0 < h_p(W) \le \frac{2M_1}{\|W\|^{2n-2}} \quad \text{on} \quad D_p \cap \{W : \|W\| = 1\}$$

which implies, by the maximum principle, that the same inequality holds on $D_p \cap \{W : \|W\| > 1\}$. Thus

$$0 < h(W) \le \frac{2M_1}{\|W\|^{2n-2}} \quad \text{on} \quad \tilde{D} \cap \{W : \|W\| > 1\}$$

and $h(W)$ is a minimal function with support $0 \in \tilde{D}$ (cf. [H], ch. 12) so that

$$h(W) = \alpha \frac{Y_{2n}}{\|W\|^{2n}}$$

for some constant $\alpha < 0$. Since $W_0 \in \partial\tilde{D} \cap \{W : \|W\| = 1\}$, it follows from Proposition 5.1 that

$$\overline{\lim_{p\to+\infty}} \frac{\left|\frac{\partial^2 G}{\partial x_i \partial x_j}(\xi_p, z_p)\right|}{\|\mathrm{Grad}_{(z)} G(\xi_p, z_p)\|} \cdot \|\xi_p - z_p\|$$

$$= \overline{\lim_{p\to+\infty}} \frac{\left|\frac{\partial^2 h_p}{\partial Y_i \partial Y_j}(W_p)\right|}{\|\mathrm{Grad}_{(W)} h_p(W_p)\|} \cdot \frac{\|\xi_p - z_p\|}{k_p}$$

$$= \frac{\left|\frac{\partial^2 h}{\partial Y_i \partial Y_j}(W_0)\right|}{\|\mathrm{Grad}_{(W)} h(W_0)\|}$$

which is finite. This contradicts (5.29); hence $0 \notin \partial\tilde{D}$.

Thus $0 \notin \tilde{D} \cup \partial\tilde{D}$ and we have reached a contradiction. Therefore (5.27) holds and the proof of Step 3 is complete. □

In order to apply the potential theory lemma, Lemma 5.1, we need to modify F on the sets $E_r(\xi, w_0)$. Let $\mathcal{E}_r(U)$ be the set on which the estimates (5.21) hold from Step 3. Let $\xi \in U$ and $w_0 \in \partial D(\xi)$. For simplicity, we write

$$E = E_r(\xi, w_0) = \{w \in D(\xi) : \|w - w_0\| < r\|w_0\|\} \equiv D(\xi) \cap B_r(w_0).$$

Step 4. There exists a function $F^*(w)$ (depending on the parameters ξ, w_0) of class C^2 on E such that

(5.33) $$H^E_{F^*}(w) \text{ (= Dirichlet solution in } E \text{ with boundary values } F^*) = \frac{\partial g}{\partial \xi_\nu}(\xi, w) \quad \text{in } E.$$

Furthermore, there exists a constant $A > 0$ independent of $\xi \in U$ and $w_0 \in \partial D(\xi)$ such that

(5.34)
$$(1)\quad |F^*(w)| < \frac{A}{\|w_0\|^{2n-3}} \quad \text{in } E;$$
$$(2)\quad \left|\frac{\partial F^*}{\partial y_i}(w_0)\right| < \frac{A}{\|w_0\|^{2n-2}} \quad i = 1, \cdots, n;$$
$$(3)\quad |\Delta_{(w)} F^*(w)| < \frac{A}{\|w_0\|^{2n-1}} \quad \text{in } E.$$

Proof of Step 4. Recall that

(5.35) $$\frac{\partial g}{\partial \xi_\nu}(\xi, w) = H^{D(\xi)}_{F(\xi,w)}(w) \quad (= H_F \text{ for short})$$

for $\xi \in \mathbb{C}^n$ and $w \in D(\xi)$. Fix $\xi \in U$ and $w_0 \in \partial D(\xi)$ and consider the harmonic function $u(w)$ in E with boundary values

$$u(w) = \begin{cases} 0 & \text{if } w \in \partial D(\xi) \cap B_r(w_0) \\ H_F - F & \text{if } w \in D(\xi) \cap \partial B_r(w_0) \end{cases}$$

Set $F^{\bullet}(w) = F(\xi,w) + u(w)$ in E. Then

$$H^E_{F^{\bullet}}(w) = H^E_F(w) + u(w)$$

is a harmonic function in E with boundary values

$$\begin{cases} F & \text{if } w \in \partial D(\xi) \cap B_r(w_0) \\ H_F & \text{if } w \in D(\xi) \cap \partial B_r(w_0) \end{cases}$$

Thus (5.35) implies that $H^E_{F^{\bullet}} = \dfrac{\partial g}{\partial \xi_\nu}$ in E, which is (5.33).

To prove the estimates (1)-(3) in (5.34), we first show that there exists $M > 0$ independent of ξ and w_0 such that

(5.36) $$|H^E_F(w)| < \frac{M}{\|w\|^{2n-3}} \quad \text{in } E.$$

By (5.21) (1) (Step 3), we have

$$|F(\xi,w)| < \frac{M}{\|w\|^{2n-3}}$$

for $(\xi,w) \in \mathcal{E}_r(U)$ and hence for $w \in E$. Since $k(w) \equiv \dfrac{1}{\|w\|^{2n-3}}$ is superharmonic in $\mathbb{C}^n$ and $H^E_F(w)$ is harmonic in E, (5.36) follows. Hence

$$|H^E_F(w) - F(\xi,w)| < \frac{2M}{\|w\|^{2n-3}} \quad \text{in } E,$$

so that, again using superharmoncity of $k(w)$, the harmonic function u satisfies the estimate

(5.37) $$|u(w)| < \frac{2M}{\|w\|^{2n-3}} \quad \text{in } E.$$

Since $E \subset \{w \in \mathbb{C}^n : \|w - w_0\| < r\|w_0\|\} = B_r(w_0)$,

$$|F^{\bullet}(w)| \le |F(\xi,w)| + |u(w)| < \frac{3M}{\|w\|^{2n-3}} < \frac{3M}{(1-r)^{2n-3}} \frac{1}{\|w_0\|^{2n-3}} \quad \text{in } E,$$

which proves (1). Note the same estimate applied to (5.37) yields that

$$|u(w)| < \frac{2M}{(1-r)^{2n-3}} \frac{1}{\|w_0\|^{2n-3}} \equiv \frac{M_1}{\|w_0\|^{2n-3}} \quad \text{in } E.$$

Thus $u(w) \le \dfrac{M_1}{\|w_0\|^{2n-3}}$ in E and $u(w) = 0$ for $w \in \partial D(\xi) \cap B_r(w_0)$. We want to apply equation (4.1) of Step 2, Chapter 4 to estimate $\|\mathrm{Grad}_{(w)} u(w_0)\|$. Recalling from Step 3 of Chapter 4 (cf. equation (4.3)) that we can find a ball $B \subset \mathbb{C}^n - D(\xi)$ with radius $\frac{\tilde{\rho}}{\ell}\|w_0\|$ which is tangent to $\partial D(\xi)$ at w_0 (with $\tilde{\rho}, \ell$ independent of ξ, w_0), equation (4.1) yields the estimate

$$\|\mathrm{Grad}_{(w)} u(w_0)\| < c\left(\frac{M_1}{\|w_0\|^{2n-3}}\right)\Big/\min(r\|w_0\|, \tfrac{\tilde{\rho}}{\ell}\|w_0\|)$$

$$= \frac{cM_1}{\min(r, \frac{\tilde{\rho}}{\ell})} \frac{1}{\|w_0\|^{2n-2}} \equiv \frac{M_2}{\|w_0\|^{2n-2}}$$

where M_2 is independent of $\xi \in U$ and $w_0 \in \partial D(\xi)$. Thus by (5.21)(2) of Step 3,

$$\left|\frac{\partial F^*}{\partial y_i}(w_0)\right| \le \left|\frac{\partial F}{\partial y_i}(w_0)\right| + \left|\frac{\partial u}{\partial y_i}(w_0)\right|$$

$$\le \frac{M + M_2}{\|w_0\|^{2n-2}}$$

which proves (2).

Finally, using the fact that u is harmonic together with (5.21)(3) of Step 3, we obtain, for $\xi \in U$,

$$|\Delta_{(w)} F^*(w)| = |\Delta_{(w)} F(\xi, w)|$$

$$\le \frac{nM}{\|w\|^{2n-1}} \qquad \text{in } E_r(\xi)$$

$$\le \frac{nM}{(1-r)^{2n-1}} \frac{1}{\|w_0\|^{2n-1}} \qquad \text{in } E$$

since $E \subset B_r(w_0)$. Thus (3) and Step 4 are proved. □

For convenience, we restate Lemma 5.1 in the precise form in which it will be used.

Lemma 5.1′. Let B_1, B_2 be two concentric balls in $\mathbb{C}^n$ with center 0 and radii ρ_1, ρ_2. Let E be a domain with smooth boundary such that $E \subset B_2 - B_1$ and ∂B_1 is tangent to ∂E at w_0. If $h(w)$ is a C^2 function in $\bar{E}$ satisfying the estimates

(5.38) (1) $\left|\frac{\partial h}{\partial y_i}(w_0)\right| < 1,$

$$(2) \quad \left|\frac{\partial^2 h}{\partial y_i \partial y_j}(w)\right| < 1 \quad \text{in } E \quad (i,j = 1,\cdots,2n),$$

then there exists a constant M, depending only on ρ_1 and ρ_2, such that

$$(5.39) \qquad \|\mathrm{Grad}_{(w)}\, H_h^E(w_0)\| \le M.$$

Proof. This follows directly from equation (5.6)(2) and the estimates (5.38)(2) and (3). □

We now can prove our required estimate on $\dfrac{\partial^2 g}{\partial \xi_\nu \partial \bar{w}_\alpha}$.

Step 5. Let $\mathcal{E}_r(U)$ be the set from Step 3. There exists a constant $c > 0$ such that

$$(5.40) \qquad \left|\frac{\partial^2 g}{\partial \xi_\nu \partial \bar{w}_\alpha}(\xi, w)\right| < \frac{c}{\|w\|^{2n-2}}$$

for all $\xi \in U$ and $w \in \overline{D(\xi)}$.

Proof of Step 5. By the maximum principle, it suffices to prove (5.40) for $\xi \in U$ and $w_0 \in \partial D(\xi)$. Given such ξ, w_0, we let $F^{\bullet}$ be a C^2 function in

$E = E_r(\xi, w_0)$ satisfying (5.33) and (5.34) from Step 4. To transform $B_r(w_0)$ to the unit ball, we consider the similarity mapping

$$\tilde{w} = T(w) = \frac{w - w_0}{r\|w_0\|}$$

and set $\tilde{E} = T(E) \subset \{\tilde{w} : \|\tilde{w}\| < 1\}$. Call

$$u(\tilde{w}) \equiv \frac{\partial g}{\partial \xi_\nu}(\xi, w) \quad \text{and} \quad h(\tilde{w}) \equiv F^{\bullet}(w).$$

In terms of the $\tilde{w}$ variables, equations (5.33) and (5.34) become

$$\text{(5.33)}' \qquad u(\tilde{w}) = H^{\tilde{E}}_{\tilde{h}}(\tilde{w})$$

and

$$\text{(1)} \quad |h(\tilde{w})| < \frac{A}{\|w_0\|^{2n-3}} \quad \text{in } \tilde{E};$$

$$\text{(5.34)}' \quad \text{(2)} \quad \left|\frac{\partial h}{\partial \tilde{y}_i}(0)\right| = \left|\frac{\partial F^{\bullet}}{\partial y_i}(w_0)\right| r\|w_0\| < \frac{Ar}{\|w_0\|^{2n-3}};$$

$$\text{(3)} \quad |\Delta_{(\tilde{w})} h(\tilde{w})| = |\Delta_{(w)} F^{\bullet}(w)| \, r^2\|w_0\|^2 \le \frac{Ar^2}{\|w_0\|^{2n-3}} \quad \text{in } \tilde{E}.$$

Thus, setting $A_1 = \max(Ar, Ar^2)$, each quantity on the left-hand side of (2); (3) in (5.34)′ is majorized by $\dfrac{A_1}{\|w_0\|^{2n-3}}$ so that we can apply Lemma 5.1′. Precisely, recalling that we can find a ball $B \subset \mathbb{C}^n - D(\xi)$ with radius $\frac{\tilde{\rho}}{\ell}\|w_0\|$ which is tangent to $\partial D(\xi)$ at w_0, setting $T(B) = \tilde{B}$, we see that the ball $\tilde{B} \subset \mathbb{C}^n - \tilde{E}$ has radius $\frac{\tilde{\rho}}{\ell}\frac{1}{r}$ and is tangent to $\partial\tilde{E}$ at 0. Since $\tilde{E} \subset \{\tilde{w} : \|\tilde{w}\| < 1\}$, we can apply Lemma 5.1′ to $\tilde{E}$, $\rho_1 = \frac{\tilde{\rho}}{\ell}\frac{1}{r}$ and $\rho_2 = \rho_1 + 2$ to get

$$\|\operatorname{Grad}_{(\tilde{w})} u(0)\| \le \frac{MA_1}{\|w_0\|^{2n-3}}$$

where M is a constant depending only on ρ_1 and ρ_2. Since

$$\frac{\partial u}{\partial \bar{\tilde{w}}_\alpha}(0) = \frac{\partial^2 g}{\partial \bar{w}_\alpha \partial \xi_\nu}(\xi, w_0) \cdot r\|w_0\| ,$$

$$\left|\frac{\partial^2 g}{\partial \bar{w}_\alpha \partial \xi_\nu}(\xi, w_0)\right| \le \left(\frac{MA_1}{r}\right) \frac{1}{\|w_0\|^{2n-2}}$$

where $\frac{MA_1}{r}$ is independent of $\xi \in U$ and $w_0 \in \partial D(\xi)$. □

<u>Step 6</u>. (B) holds, i.e., given $\xi_0 \in \partial D$,

$$\text{(B)} \qquad \lim_{R\to+\infty} \int_{\partial D(\xi)\cap\{w:\,\|w\| > R\}} \frac{\frac{\partial f}{\partial \xi_\nu}}{\|\operatorname{Grad}_{(w)} f\|} \frac{\partial^2 g}{\partial \bar{\xi}_\nu \partial w_\alpha} \frac{\partial g}{\partial n_w} dS_w = 0$$

uniformly for ξ in a neighborhood of $\xi_0 \in \mathbb{C}^n$.

<u>Proof of Step 6</u>. Fix a neighborhood U of ∂D so that (5.40) holds. Then for any $\xi \in U$, $R > 0$, using the estimates

$$\left|\frac{\partial f}{\partial \xi_\nu}(\xi, w)\right| \le A\|w\|^2, \quad \|\operatorname{Grad}_{(w)} f(\xi, w)\| \ge m > 0$$

from Step 4, Chapter 4, we obtain

$$\int_{\partial D(\xi)\cap\{w:\,\|w\| > R\}} \frac{\left|\frac{\partial f}{\partial \xi_\nu}\right|}{\|\operatorname{Grad}_{(w)} f\|} \left|\frac{\partial^2 g}{\partial \bar{\xi}_\nu \partial w_\alpha}\right| \left(\frac{-\partial g}{\partial n_w}\right) dS_w$$

$$< \int_{\partial D(\xi)\cap\{w:\,\|w\| > R\}} \frac{A\|w\|^2}{m} \frac{M}{\|w\|^{2n-2}} \left(\frac{-\partial g}{\partial n_w}\right) dS_w$$

$$\le \frac{AM}{m} \frac{1}{R^{2n-4}} \int_{\partial D(\xi)\cap\{w:\,\|w\| > R\}} \left(\frac{-\partial g}{\partial n_w}\right) dS_w$$

Given $\xi_0 \in \partial D$ and $\varepsilon > 0$, we can find a small ball

$$b = \{\xi : \|\xi - \xi_0\| < \rho\} \subset\subset U$$

and $R >> 1$ such that

$$\int_{\partial D(\xi)\cap\{w: \|w\| > R\}} \left(\frac{-\partial g}{\partial n_w}\right) dS_w < \varepsilon$$

for all $\xi \in b$. Since $n \ge 2$, we get that

$$\int_{\partial D(\xi)\cap\{w: \|w\| > R\}} |F(\xi,w)| \left|\frac{\partial^2 g}{\partial\bar{\xi}_\nu \partial w_\alpha}\right| dS_w < \frac{AM}{2m}\varepsilon$$

for all $\xi \in b$, including ξ_0, which proves Step 6. □

To prove that (A) holds, i.e., the above integrals over $\partial D(\xi) \cap \{w : \|w\| < R\}$ converge as $\xi \to \xi_0$ for any $R > 1$, it suffices to show the following improvement of Step 6, Chapter 4.

<u>Step</u> <u>7</u>. (1) $\frac{\partial g}{\partial \xi_\nu}(\xi,w)$ is continuous for $(\xi,w) \in \bigcup_{\xi\in\mathbb{C}^n} (\xi, \overline{D(\xi)})$

(2) $\frac{\partial^2 g}{\partial \xi_\nu \partial \bar{w}_\alpha}(\xi,w)$ is continuous for $(\xi,w) \in \bigcup_{\xi\in\mathbb{C}^n} (\xi, \overline{D(\xi)})$.

<u>Proof</u> <u>of</u> <u>Step</u> <u>7</u>. From the results of Chapter 4, it suffices to prove continuity at $(\xi_0,w_0) \in \partial D \times \partial D(\xi_0)$. Again let $\mathcal{E}_r(U)$ be the set from Step 2 and choose $\eta, r^* > 0$ so that the balls

$$b \equiv \{\xi : \|\xi - \xi_0\| < \eta\} \subset\subset U$$

and

$$B^* \equiv \{w : \|w - w_0\| < r^*\|w_0\|\}$$

satisfy

$B^{\bullet} \cap D(\xi) \subset\subset E_r(\xi)$ $(= \bigcup_{w_0 \in \partial D(\xi)} [B_r(w_0) \cap D(\xi)])$ for all $\xi \in b$

and

$$\partial D(\xi) \cap B^{\bullet} \neq \emptyset \quad \text{for all} \quad \xi \in b.$$

Letting $\tilde{\rho}, \ell$ be as in the proof of Step 5 (so that we can find a ball in $\mathbb{C}^n - D(\xi)$ with radius $\frac{\tilde{\rho}}{\ell} \|w_0\|$ tangent to $\partial D(\xi)$ at w_0), we choose ρ_2, ρ_1 satisfying

$$0 < \rho_1 < \frac{\tilde{\rho}}{\ell}(1 - r^{\bullet})\|w_0\| \quad \text{and} \quad \rho_2 > \rho_1 + 2(1 + r^{\bullet})\|w_0\|.$$

Then for each $\xi \in b$ and $w^{\bullet} \in \partial D(\xi) \cap B^{\bullet}$, we can find a ball of radius ρ_1,

$$B_1(\xi, w^{\bullet}) \equiv \{w : \|w - \alpha(\xi, w^{\bullet})\| < \rho_1\} \subset \mathbb{C}^n - D(\xi),$$

which is tangent to $\partial D(\xi)$ at $w^{\bullet}$. By the choice of ρ_2, we then have

$$B^{\bullet} \cap D(\xi) \subset B_2(\xi, w^{\bullet}) - B_1(\xi, w^{\bullet})$$

where $B_2(\xi, w^{\bullet})$ is the concentric ball

$$B_2(\xi, w^{\bullet}) \equiv \{w : \|w - \alpha(\xi, w^{\bullet})\| < \rho_2\}.$$

Clearly for each $\xi \in b$ we can choose a point $w^{\bullet} = w^{\bullet}(\xi) \in \partial D(\xi) \cap B^{\bullet}$ so that the mapping

$$\xi \longmapsto w^{\bullet}(\xi)$$

is continuous for $\xi \in b$ and $w^{\bullet}(\xi_0) = w_0$. Then the function

$$\xi \longmapsto \alpha(\xi, w^{\bullet}(\xi)) \equiv \alpha(\xi)$$

yielding the center of the balls B_1, B_2 is also continuous for $\xi \in b$. Proceeding as in the proof of Lemma 5.1, we define the function

$$s(\xi, w) \equiv -(\rho_2^2 - \rho_1^2) \left[\frac{\dfrac{1}{\rho_1^{2n-2}} - \dfrac{1}{\|w\|^{2n-2}}}{\dfrac{1}{\rho_1^{2n-2}} - \dfrac{1}{\rho_2^{2n-2}}} \right] + \|w - \alpha(\xi)\|^2 - \rho_1^2 .$$

We want to use $s(\xi, w)$ to estimate

$$\frac{\partial g}{\partial \xi_\nu}(\xi, w) = H_F^{D(\xi)}(\xi, w)$$

on $B^\bullet \cap D(\xi)$. From the remark after Step 3, we recall that there exists $r_0 > 0$ such that

$$\{w : \|w\| < r_0\} \subset\subset D(\xi) \quad \text{for all} \quad \xi \in \mathbb{C}^n ;$$

hence we can assume that r is sufficiently small to guarantee

$$[U \times \{w : \|w\| < r_0\}] \cap \mathcal{E}_r(U) = \emptyset.$$

Thus $(\xi, w) \in \mathcal{E}_r(U)$ implies that $\|w\| \geq r_0$ and hence (5.21)(1) of Step 3 yields

$$|\Delta_{(w)} F(\xi, w)| < \frac{nM}{r_0^{2n-1}} \equiv M_1 \quad \text{in} \quad \mathcal{E}_r(U).$$

Following the argument of Lemma 5.1, we obtain

$$\text{(5.41)} \qquad \operatorname{Re} F(\xi, w) + \frac{M_1}{n} s(\xi, w) \leq \operatorname{Re} \frac{\partial g}{\partial \xi_\nu}(\xi, w) \leq \operatorname{Re} F(\xi, w) - \frac{M_1}{n} s(\xi, w)$$

for $w \in B^\bullet \cap \overline{D(\xi)}$. Furthermore, since $B_1(\xi, w^\bullet)$ is tangent to $\partial D(\xi)$ at $w^\bullet(\xi)$, $s(\xi, w^\bullet(\xi)) = 0$ and equality holds in (5.41) at $w = w^\bullet(\xi)$. Since $s(\xi, w)$ is clearly continuous for

$$(\xi, w) \in \bigcup_{\xi \in b} (\xi, \overline{D(\xi)} \cap B^\bullet) \subset \mathcal{E}_r(U)$$

and $F(\xi, w)$ is continuous on the same set, it follows from (5.41) and the above remarks that

$$\lim_{(\xi,w)\to(\xi_0,w_0)} \operatorname{Re}\frac{\partial g}{\partial \xi_\nu}(\xi,w) = \operatorname{Re} F(\xi_0,w_0).$$

A similar argument shows that

$$\lim_{(\xi,w)\to(\xi_0,w_0)} \operatorname{Im}\frac{\partial g}{\partial \xi_\nu}(\xi,w) = \operatorname{Im} F(\xi_0,w_0) ;$$

hence $\frac{\partial g}{\partial \xi_\nu}(\xi,w)$ is continuous at (ξ_0,w_0) and (1) is proved.

To prove (2), we fix $\xi \in b$ and write

$$\frac{\partial g}{\partial \xi_\nu}(\xi,w) = h_1(\xi,w) + h_2(\xi,w) \quad \text{for} \quad w \in B^* \cap D(\xi)$$

where h_1, h_2 are harmonic functions of $w \in B^* \cap D(\xi)$ with boundary values

$$h_1(\xi,w) = \begin{cases} \frac{\partial g}{\partial \xi_\nu}(\xi,w) & \text{if} \quad w \in \partial B^* \cap D(\xi) \\ 0 & \text{if} \quad w \in \partial D(\xi) \cap B^* \end{cases}$$

and

$$h_2(\xi,w) = \begin{cases} 0 & \text{if} \quad w \in \partial B^* \cap D(\xi) \\ F(\xi,w) & \text{if} \quad w \in \partial D(\xi) \cap B^* \end{cases}$$

From (1) of Step 7, it follows that $h_1(\xi,w)$ is continuous for $(\xi,w) \in \bigcup_{\xi\in b} (\xi, \overline{B^* \cap D(\xi)})$ except perhaps at the corners $\bigcup_{\xi\in b} (\xi, \partial B^* \cap \partial D(\xi))$. Since $h_1(\xi,w) = 0$ for $w \in \partial D(\xi) \cap B^*$, it follows from Preliminary 4.2 [Y] that $\frac{\partial h_1}{\partial \bar{w}_\alpha}(\xi,w)$ is continuous at (ξ_0,w_0). To verify the continuity of $\frac{\partial h_2}{\partial \bar{w}_\alpha}(\xi,w)$ at (ξ_0,w_0), we just remark that this can be proved from the concrete construction of harmonic functions via the theory of Fredholm integral equations together with the fact that $F(\xi,w)$ is of class C^4 in $\bigcup_{\xi\in b}(\xi, \overline{B^{\bullet}\cap D(\xi)})$. Thus (2) is proved. □

We now complete the proof of Lemma 3.1.

<u>Step 8</u>. $\lambda(\xi)$ is of class C^2 in $\mathbb{C}^n$.

<u>Proof of Step 8</u>. We have shown that for $\xi_0 \in \partial D$,

$$\lim_{\xi\to\xi_0} \frac{\partial^2\lambda}{\partial\xi_\nu\partial\bar{\xi}_\nu}(\xi), \quad \nu = 1,\cdots,n,$$

exists. By a unitary change of coordinates, it follows that

$$\lim_{\xi\to\xi_0} \frac{\partial^2\lambda}{\partial\xi_\nu\partial\bar{\xi}_\mu}(\xi), \quad \mu,\nu = 1,\cdots,n,$$

exists. To complete the proof of Step 8, we must show that

$$\lim_{\xi\to\xi_0} \frac{\partial^2\lambda}{\partial y_i\partial y_j}(\xi), \quad i,j = 1,\cdots,2n \tag{5.42}$$

exists where $\xi_k = y_{2k-1} + iy_{2k}$, $k = 1,\cdots,n$.

In order to prove (5.42), it is necessary to reconstruct the arguments used in this paper in the real category, i.e., using variations of domains in R^{2n} with parameter space $I \subset R$. For example, for $t \in I \subset R$ and $x \in R$, we define the <u>real Levi-form of ψ with respect to (t,x)</u> for a C^2 function ψ to be

$$L_{(t,x)}\psi = \left|\frac{\partial\psi}{\partial t}\right|^2 \frac{\partial^2\psi}{\partial x^2} - 2\,\frac{\partial\psi}{\partial t}\,\frac{\partial\psi}{\partial x}\,\frac{\partial^2\psi}{\partial t\partial x} + \frac{\partial^2\psi}{\partial t^2}\left|\frac{\partial\psi}{\partial x}\right|^2 .$$

Then we can define $\mathcal{L}\psi$ for ψ defined on a domain in $I \times R^{2n}$ in a manner analogous to (1.2) and develop fundamental formulas for derivatives of the Green function and Robin function similar to the complex case (c.f. [Y], Chapter 9). Proceeding in this manner, we obtain (5.42). Following the argument in Step 6, Chapter 4, it follows that formula (5.1) holds for all $\xi \in \mathbb{C}^n$ and $\lambda(\xi)$ is of class C^2 in $\mathbb{C}^n$. ∎

Remark. We do not know if Lemma 3.1 is the best possible result concerning smoothness of the Robin function $\lambda(\xi)$ up to ∂D.

As noted in the beginning of this chapter, the same reasoning as in the proof of Lemma 3.1 yields the following result.

Lemma 3.2. $\frac{\partial g}{\partial \xi_\nu}(\xi, w)$ is of class C^1 for $(\xi, w) \in \bigcup_{\xi \in \mathbb{C}^n} (\xi, D(\xi))$ and (4.5), (4.6) hold for all $\xi \in \mathbb{C}^n$.

In the proof of Lemma 3.1, we have used the formula (3.4) for $f(\xi, w)$ under the condition that the domain $D \subset \mathbb{C}^n$ is defined by a double $(\tilde{D}, \psi)$ where $\tilde{D} = \mathbb{C}^n$. In the case where D is an unramified domain over $\mathbb{C}^n$, we cannot take $\tilde{D} = \mathbb{C}^n$; thus we make a small modification.

We let $(\tilde{D}, \psi)$ be a double defining D and we set $D' = \tilde{D} - \bar{D}$. Again, for $\xi \in D$ (D') we let $G(\xi, z)$ and $\Lambda(\xi)$ denote the Green function and Robin constant for (D, ξ) $((D', \xi))$ and set

$$T_\xi(z) = w = \frac{z-\xi}{-\psi(\xi)} \quad \text{for} \quad \xi \in D \cup D'$$

and

$$D(\xi) = \begin{cases} T_\xi(D) & \text{if} \quad \xi \in D \\ \{w \in \mathbb{C}^n : 2 \operatorname{Re} \{ \sum_{\alpha=1}^{n} \frac{\partial \psi}{\partial \xi_\alpha}(\xi) w_\alpha \} - 1 < 0 \} & \text{if} \quad \xi \in \partial D . \\ T_\xi(D') & \text{if} \quad \xi \in D' \end{cases}$$

Since $0 \in D(\xi)$, $\forall\ \xi \in \tilde{D}$, we have the Green functions $g(\xi,w)$ and Robin constants $\lambda(\xi)$ for $(D(\xi),0)$ which satisfy

$$g(\xi,w) = \psi(\xi)^{2n-2}\ G(\xi,z)$$

and

$$\lambda(\xi) = \psi(\xi)^{2n-2}\ \Lambda(\xi)$$

if $\xi \in D \cup D'$, $w \in D(\xi)$; for $\xi \in \partial D$, we have

$$\begin{cases} g(\xi,w) = \dfrac{1}{\|w\|^{2n-2}} - \dfrac{1}{\|w - \bar{N}_\xi\|^{2n-2}} \\ \lambda(\xi) = -\ \|\mathrm{Grad}\ \psi(\xi)\|^{2n-2} \end{cases}$$

where $N_\xi = \dfrac{\mathrm{Grad}\ \psi(\xi)}{\|\mathrm{Grad}\ \psi(\xi)\|^2}$. In this setting, we let

$$\tilde{D}(\xi) \equiv \begin{cases} T_\xi(\tilde{D}) & \text{if } \xi \in D \cup D' \\ \mathbb{C}^n & \text{if } \xi \in \partial D \end{cases} .$$

The reason for this definition is that for any fixed $R > 1$, $\tilde{D}(\xi)$ contains the ball $\{w : \|w\| < R\}$ for $\xi \in D \cup D'$ sufficiently close to ∂D. We now write

$$\mathcal{D} \equiv \bigcup_{\xi\in\tilde{D}} (\xi, D(\xi)) \quad \text{and} \quad \tilde{\mathcal{D}} \equiv \bigcup_{\xi\in\tilde{D}} (\xi,\tilde{D}(\xi)).$$

Again we set $\tilde{\psi}(\xi,w) = \psi(\xi - \psi(\xi)w)$ for $(\xi,w) \in \tilde{\mathcal{D}}$ and we see that there exists a C^∞ function $f(\xi,w)$ in $\tilde{\mathcal{D}}$ such that $\tilde{\psi}(\xi,w) = -f(\xi,w)\psi(\xi)$. The integral formula (3.4) is no longer valid for all $(\xi,w) \in \tilde{\mathcal{D}}$ since $\tilde{\mathcal{D}}$ is not necessarily convex. However, for each fixed $\xi_0 \in \partial D$ and $R > 1$, there exists a ball $b = \{\xi : \|\xi - \xi_0\| < \rho\} \subset \tilde{D}$ with

$$b \times \{w : \|w\| < R\} \subset \tilde{\mathcal{D}}$$

such that the formula

$$f(\xi,w) = 2 \operatorname{Re}\left\{\sum_{\alpha=1}^{n} \int_0^1 w_\alpha \frac{\partial\psi}{\partial z_\alpha}\Big|_{z=\xi-\psi(\xi)tw} dt\right\} - 1$$

is valid for $(\xi,w) \in b \times \{w : \|w\| < R\}$. Thus

$$\mathcal{D} = \{(\xi,w) \in \tilde{\mathcal{D}} : f(\xi,w) < 0\} ,$$

$$D(\xi) = \{w \in \tilde{D}(\xi) : f(\xi,w) < 0\},$$

$$\operatorname{Grad}_{(w)} f(\xi,w) \neq (0,\cdots,0) \quad \text{for} \quad w \in \partial D(\xi),$$

and, by a similar argument as in the rest of this chapter, we have the following modification of Lemma 3.1.

Lemma 3.1′. Under the above hypotheses on $\mathcal{D}$, $\tilde{\mathcal{D}}$, ψ,

(1) $g(\xi,w)$ is of class C^2 in $\tilde{\mathcal{D}} \cup \partial\tilde{\mathcal{D}} - \tilde{\mathcal{D}} \times \{0\}$ and $\lambda(\xi)$ is of class C^2 on $\tilde{D}$.

(2) $\frac{\partial g}{\partial \xi_\alpha}(\xi,w)$, $\frac{\partial^2 g}{\partial\bar{\xi}_\alpha \partial \xi_\alpha}(\xi,w)$ are continuous functions in $\tilde{\mathcal{D}}$ and

$$\frac{\partial g}{\partial \xi_\alpha}(\xi,a) = \frac{1}{2(n-1)w_{2n}} \int_{\partial D(\xi)} K_1(\xi,w)\|\operatorname{Grad}_{(w)} g(\xi,w)\| \frac{\partial g_a(\xi,w)}{\partial n_w} dS_w ;$$

(5.43)

$$\frac{\partial^2 g}{\partial\bar{\xi}_\alpha \partial\xi_\alpha}(\xi,a) = \frac{1}{2(n-1)w_{2n}} \int_{\partial D(\xi)} K_2(\xi,w)\|\operatorname{Grad}_{(w)} g(\xi,w)\| \frac{\partial g_a(\xi,w)}{\partial n_w} dS_w$$

$$- \frac{4}{(n-1)w_{2n}} \operatorname{Re} \iint_{\partial D(\xi)} \sum_{i=1}^{n} \left\{\frac{\partial^2 g}{\partial\xi_\alpha \partial\bar{w}_i} \frac{\partial^2 g_a}{\partial\bar{\xi}_\alpha \partial w_i}\right\} dV_w$$

for $\xi \in \tilde{D}$ and $a \in D(\xi)$;

$$(5.44)\qquad \frac{\partial\lambda}{\partial\xi_\alpha}(\xi) = \frac{\partial g}{\partial\xi_\alpha}(\xi,0); \quad \frac{\partial^2\lambda}{\partial\xi_\alpha\partial\bar{\xi}_\alpha}(\xi) = \frac{\partial^2 g}{\partial\xi_\alpha\partial\bar{\xi}_\alpha}(\xi,0),$$

where

$$K_1 = K_1^{(\alpha)} = \frac{\dfrac{\partial f}{\partial\xi_\alpha}}{\|\mathrm{Grad}_{(w)} f\|} \quad \text{and} \quad K_2 = \frac{\mathcal{L}^{(\alpha)} f}{\|\mathrm{Grad}_{(w)} f\|^3}$$

(cf., (4.7) and (4.8)).

We write $\mathcal{D}_D \equiv \bigcup_{\xi\in D} (\xi,\ D(\xi))$ and $\mathcal{D}_{\bar{D}} \equiv \bigcup_{\xi\in\bar{D}} (\xi,\ D(\xi))$. Then $\mathcal{D}_D$ is a smooth variation of domains and Lemma 3.1′ says that $g(\xi,w)$ and $\lambda(\xi)$ are C^2 for ξ up to ∂D.

6. LIMITING FORMULAS

We now study the relationship between the Green functions $G(\xi,z)$ and $g(\xi,w)$ and the limiting behavior of the Robin function $\Lambda(\xi)$ as $\xi \to \xi_0 \in \partial D$ more precisely. In this chapter, as in Chapter 3, we assume that D is a domain in $\mathbb{C}^n$ with smooth boundary ∂D defined by a double $(\mathbb{C}^n,\psi)$. Since the variation $\mathcal{D}$: $\xi \to D(\xi)$, $\xi \in \mathbb{C}^n$, is defined by the C^∞ function $f(\xi,w)$ and since $0 \in D(\xi)$, we remark, for future use, that there exists $r_0 > 0$ such that

$$\text{(6.1)} \qquad \{w:\ \|w\| < r_0\} \subset\subset D(\xi) \quad \text{for all} \quad \xi \in \bar{D}\,.$$

We want to discuss derivatives of the Green function $G(\xi,z)$ for (D,ξ) and relate these to derivatives of the Green function $g(\xi,w)$ for $(D(\xi),0)$. Recall that

$$\text{(6.2)} \qquad G(\xi,z) = \frac{1}{\|z-\xi\|^{2n-2}} + \Lambda(\xi) + H(\xi,z)$$

where $H(\xi,z)$ is harmonic for $z \in D$ and

$$\text{(6.3)} \qquad H(\xi,\xi) = 0\ ;$$

the relationship with $g(\xi,w)$ is that

$$\text{(6.4)} \qquad g(\xi,w) = \psi(\xi)^{2n-2}\, G(\xi,z)$$

for $\xi \in D$ where $w = \dfrac{z-\xi}{-\psi(\xi)}$. It is well-known that $G(\xi,z)$ is real-analytic in $D \times D$ and satisfies $G(\xi,z) = G(z,\xi)$. We define the functions

$$\text{(6.5)} \quad G_\alpha(\xi,z) \equiv \left(\frac{\partial G}{\partial \xi_\alpha} + \frac{\partial G}{\partial z_\alpha}\right)(\xi,z) \quad \text{for} \quad (\xi,z) \in D \times D\,,\ \alpha = 1,\ldots,n$$

and

$$(6.6)\quad G_{\alpha\bar{\beta}}(\xi,z) \equiv \left(\frac{\partial G_\alpha}{\partial\bar{\xi}_\beta} + \frac{\partial G_\alpha}{\partial\bar{z}_\beta}\right)(\xi,z) \quad \text{for} \quad (\xi,z) \in D \times D\ ,\ \alpha,\beta = 1,\dots,n\ .$$

<u>Proposition</u> <u>6.1</u>. $G_\alpha, G_{\alpha\bar{\beta}}$ are real analytic, symmetric functions in $D \times D$ which are harmonic in z and in ξ and satisfy

(1) $$\frac{\partial\Lambda}{\partial\xi_\alpha}(\xi) = G_\alpha(\xi,\xi) \quad \text{and}$$

(2) $$\frac{\partial^2\Lambda}{\partial\xi_\alpha\partial\bar{\xi}_\beta}(\xi) = G_{\alpha\bar{\beta}}(\xi,\xi) = 2\,\frac{\partial G_\alpha}{\partial\bar{z}_\beta}(\xi,\xi) = 2\,\frac{\partial G_\alpha}{\partial\bar{\xi}_\beta}(\xi,\xi)\ .$$

<u>Proof</u>. Since $v(\xi,z) \equiv G(\xi,z) - \dfrac{1}{\|z-\xi\|^{2n-2}}$ is real analytic, symmetric and harmonic in $D \times D$, if we use the fact that $\left(\dfrac{\partial}{\partial\xi_\alpha} + \dfrac{\partial}{\partial z_\alpha}\right)\dfrac{1}{\|z-\xi\|^{2n-2}} \equiv 0$, it follows that $G_\alpha(\xi,z)$ has no singularity, even along the diagonal $z = \xi$. Thus G_α and $G_{\alpha\bar{\beta}}$ are harmonic, symmetric functions in $D \times D$. To prove (1) and (2) , we use (6.2) and (6.3) to obtain

$$G_\alpha(\xi,\xi) = \frac{\partial\Lambda}{\partial\xi_\alpha}(\xi) + \frac{\partial H}{\partial\xi_\alpha}(\xi,\xi) + \frac{\partial H}{\partial z_\alpha}(\xi,\xi) = \frac{\partial\Lambda}{\partial\xi_\alpha}(\xi)\ ,$$

which is (1) . Differentiating this equation with respect to $\bar{\xi}_\beta$, we obtain

$$\frac{\partial^2\Lambda}{\partial\bar{\xi}_\beta\partial\xi_\alpha}(\xi) = \frac{\partial G_\alpha}{\partial\bar{\xi}_\beta}(\xi,\xi) + \frac{\partial G_\alpha}{\partial\bar{z}_\beta}(\xi,\xi) \equiv G_{\alpha\bar{\beta}}(\xi,\xi)\ .$$

To complete the proof of (2) , we differentiate the symmetry formula $G_\alpha(\xi,z) = G_\alpha(z,\xi)$ with respect to $\bar{\xi}_\beta$ and set $z = \xi$ to obtain $\dfrac{\partial G_\alpha}{\partial\bar{z}_\beta}(\xi,\xi) = \dfrac{\partial G_\alpha}{\partial\bar{\xi}_\beta}(\xi,\xi)$ which gives (2) . ∎

Related to $G_\alpha(\xi,z)$, we define, for $(\xi,w) \in \mathcal{D}_{\bar{D}} = \bigcup_{\xi\in\bar{D}} (\xi,D(\xi))$,

$$(6.7)\quad \begin{cases} g_0(\xi,w) \equiv g(\xi,w) + \dfrac{1}{n-1} \displaystyle\sum_{i=1}^{n} w_i \frac{\partial g}{\partial w_i}(\xi,w) \quad \text{and} \\[2ex] g_\alpha(\xi,w) \equiv \psi(\xi) \dfrac{\partial g}{\partial \xi_\alpha}(\xi,w) - (n-1) \dfrac{\partial \psi}{\partial \xi_\alpha}(\xi) \,[g_0(\xi,w) + \overline{g_0(\xi,w)}] , \\ \qquad \alpha = 1,\dots,n . \end{cases}$$

Note that by Lemma 3.2 , these are well-defined functions on $\mathcal{D}_{\bar{D}}$.

Proposition 6.2. (in all that follows, $\alpha = 1,\dots,n$)

(1) For $\xi \in \bar{D}$, $g_0(\xi,w)$ and $g_\alpha(\xi,w)$ are harmonic functions of $w \in D(\xi)$;

(2) g_0, g_α are continuous for $(\xi,w) \in \mathcal{D}_{\bar{D}}$;

(3) $g_0(\xi,0) = \lambda(\xi)$ for $\xi \in \bar{D}$;

(6.8) (4) $g_\alpha(\xi,0) = \psi(\xi) \dfrac{\partial g}{\partial \xi_\alpha}(\xi,0) - (2n-2) \dfrac{\partial \psi}{\partial \xi_\alpha}(\xi)\, \lambda(\xi)$ for $\xi \in \bar{D}$;

(6.9) (5) $g_\alpha(\xi,w) = G_\alpha(\xi,z)\psi(\xi)^{2n-1}$ for $(\xi,w) \in \mathcal{D}_D$ where $w = \dfrac{z-\xi}{-\psi(\xi)}$.

Proof. For $w \neq 0$,

$$\text{(i)}\quad \sum_{i=1}^{n} w_i \frac{\partial}{\partial w_i}\left(\frac{1}{\|w\|^{2n-2}} \right) = -(n-1)\frac{1}{\|w\|^{2n-2}} ;$$

and if u is a C^2 function ,

$$\text{(ii)}\quad \Delta_{(w)}\left(\sum_{i=1}^{n} w_i \frac{\partial u}{\partial w_i} \right) = \Delta_{(w)}u + \sum_{i=1}^{n} w_i \frac{\partial}{\partial w_i}(\Delta_{(w)}u) .$$

Using (i), the definition of $g_0(\xi,w)$ in (6.7) and the definition of $g(\xi,w)$ in (3.5) , it follows that

$g_0(\xi,w) = \lambda(\xi) + \sum_{i=1}^{n} w_i \frac{\partial h}{\partial w_i}(\xi,w)$ for $w \in D(\xi)$ if $\xi \in \bar{D}$; i.e., the singularity $\frac{1}{\|w\|^{2n-2}}$ cancels and $g_0(\xi,w)$ is a harmonic function of $w \in D(\xi)$ -- this follows from (ii) since h is harmonic. Furthermore $g_0(\xi,0) = \lambda(\xi)$ from this computation and the properties of the function $g_0(\xi,w)$ in (1) and (3) are established.

Since $\frac{\partial g}{\partial \xi_\alpha}(\xi,w)$ is harmonic for $w \in D(\xi)$ if $\xi \in \bar{D}$, the harmonicity of $g_\alpha(\xi,w)$ for $w \in D(\xi)$ follows from that of $g_0(\xi,w)$ and the definition of g_α in (6.7). Equation (4) follows from (6.7) by setting $w = 0$ and using (3). (note $\lambda(\xi)$ is real).

The continuity of g_0 in $\mathcal{D}_{\bar{D}}$ follows from the continuity of g in $\mathcal{D}_{\bar{D}} - \bar{D} \times \{0\}$ (Lemma 3.1, Step 1) ; similarly, the continuity of g_α in $\mathcal{D}_{\bar{D}}$ follows from the continuity of $\frac{\partial g}{\partial \xi_\alpha}$ in $\mathcal{D}_{\bar{D}}$ (Lemma 3.2) and (1); hence (2) is proved. Finally, to prove (5), we differentiate

$$g(\xi,w) = \psi(\xi)^{2n-2} G(\xi,z) \tag{6.4}$$

with respect to z_α and with respect to ξ_α to obtain, using $w = \frac{z-\xi}{-\psi(\xi)}$,

$$\frac{\partial g}{\partial w_\alpha}\left(\frac{1}{-\psi}\right) = \psi^{2n-2}\frac{\partial G}{\partial z_\alpha} \quad \text{and}$$

$$\frac{\partial g}{\partial \xi_\alpha} + \frac{1}{\psi}\frac{\partial g}{\partial w_\alpha} + \frac{\frac{\partial \psi}{\partial \xi_\alpha}}{\psi^2}\sum_{i=1}^{n}\left\{(z_i-\xi_i)\frac{\partial g}{\partial w_i} + (\bar{z}_i-\bar{\xi}_i)\frac{\partial g}{\partial \bar{w}_i}\right\} = (2n-2)\psi^{2n-3}\frac{\partial \psi}{\partial \xi_\alpha}G + \psi^{2n-2}\frac{\partial G}{\partial \xi_\alpha}.$$

Using $w_i = \frac{z_i-\xi_i}{-\psi(\xi)}$, $g = \psi^{2n-2}G$, and $G_\alpha \equiv \frac{\partial G}{\partial z_\alpha} + \frac{\partial G}{\partial \xi_\alpha}$, we obtain

$$\frac{\partial g}{\partial \xi_\alpha} - \frac{\frac{\partial \psi}{\partial \xi_\alpha}}{\psi} \sum_{i=1}^{n} \left(w_i \frac{\partial g}{\partial w_i} + \bar{w}_i \frac{\partial g}{\partial \bar{w}_i} \right) = (2n-2) \frac{\frac{\partial \psi}{\partial \xi_\alpha}}{\psi} g + \psi^{2n-2} G_\alpha$$

so that

$$G_\alpha(\xi, z) = \frac{1}{\psi^{2n-1}} \left[\psi \frac{\partial g}{\partial \xi_\alpha} - (n-1) \frac{\partial \psi}{\partial \xi_\alpha} \left\{ \left(g + \frac{1}{n-1} \sum_{i=1}^{n} w_i \frac{\partial g}{\partial w_i} \right) \right. \right.$$

$$\left. \left. + \left(g + \frac{1}{n-1} \sum_{i=1}^{n} \bar{w}_i \frac{\partial g}{\partial \bar{w}_i} \right) \right\} \right]$$

$$= \frac{1}{\psi^{2n-2}} g_\alpha(\xi, w) \quad \text{for} \quad \xi \in D, \ w \in D(\xi). \quad \blacksquare$$

Note that for $\xi_0 \in \partial D$, the explicit formula (4.21) for $g(\xi_0, w)$ yields

$$(6.10) \quad g_0(\xi_0, w) = \sum_{i=1}^{n} \frac{(\bar{N}_{\xi_0})_i \bar{w}_i - \|N_{\xi_0}\|^2}{\|w - \bar{N}_{\xi_0}\|^{2n}} \quad \text{where} \quad \bar{N}_{\xi_0} = \left((\bar{N}_{\xi_0})_1, \ldots, (\bar{N}_{\xi_0})_n \right).$$

These formulas (6.1) - (6.10) will be used throughout the rest of the paper.

We now use Lemma 3.1 to prove the main result in this section on boundary behavior of $\Lambda(\xi)$.

Lemma 6.1. Let D be a domain in $\mathbb{C}^n$ with smooth boundary. Let $\xi_0 \in \partial D$. Then for $\xi \in D$,

(1) $$\lim_{\xi \to \xi_0} \Lambda(\xi) \psi(\xi)^{2n-2} = - \|\text{Grad } \psi(\xi_0)\|^{2n-2}$$

(2) $$\lim_{\xi \to \xi_0} \frac{\partial \Lambda}{\partial \xi_\alpha}(\xi) \psi(\xi)^{2n-1} = (2n-2) \frac{\partial \psi}{\partial \xi_\alpha}(\xi_0) \|\text{Grad } \psi(\xi_0)\|^{2n-2}$$

(3) $$\lim_{\xi \to \xi_0} \frac{\partial^2 \Lambda}{\partial \xi_\alpha \partial \bar{\xi}_\beta}(\xi) \psi(\xi)^{2n} = -(2n-1)(2n-2) \frac{\partial \psi}{\partial \xi_\alpha}(\xi_0) \frac{\partial \psi}{\partial \bar{\xi}_\beta}(\xi_0) \cdot \|\text{Grad } \psi(\xi_0)\|^{2n-2}$$

$$(4)\quad \lim_{\xi\to\xi_0} \frac{\partial^2 \log(-\Lambda)}{\partial\xi_\alpha \partial\bar{\xi}_\beta}(\xi)\psi(\xi)^2 = (2n-2)\frac{\partial\psi}{\partial\xi_\alpha}(\xi_0)\frac{\partial\psi}{\partial\bar{\xi}_\beta}(\xi_0) .$$

<u>Proof</u>. (1) is a consequence of the continuity of the function

$$\lambda(\xi) \equiv \begin{cases} \Lambda(\xi)\psi(\xi)^{2n-2} , & \xi \in D \\ -\|\mathrm{Grad}\ \psi(\xi)\|^{2n-2}, & \xi \in \partial D \end{cases}$$

up to ∂D proved in Lemma 3.1. Since λ was shown to be C^1 up to ∂D and $\psi(\xi_0) = 0$,

$$\lim_{\xi\to\xi_0} \frac{\partial\lambda}{\partial\xi_\alpha}(\xi)\ \psi(\xi) = 0$$

so that

$$\lim_{\xi\to\xi_0} \frac{\partial\Lambda}{\partial\xi_\alpha}(\xi)\ \psi(\xi)^{2n-1} = \lim_{\xi\to\xi_0}\left[\frac{\partial\lambda}{\partial\xi_\alpha}(\xi)\psi(\xi) - (2n-2)\lambda(\xi)\frac{\partial\psi}{\partial\xi_\alpha}(\xi)\right]$$

$$= -(2n-2)\lambda(\xi_0)\frac{\partial\psi}{\partial\xi_\alpha}(\xi_0)$$

which is (2) . To prove (3), note that since

$$\frac{\partial\Lambda}{\partial\xi_\alpha}(\xi)\psi(\xi)^{2n-1} = \frac{\partial\lambda}{\partial\xi_\alpha}(\xi)\psi(\xi) - (2n-2)\lambda(\xi)\frac{\partial\psi}{\partial\xi_\alpha}(\xi) ,$$

we have

$$\frac{\partial^2\Lambda}{\partial\xi_\alpha\partial\bar{\xi}_\beta}(\xi)\psi(\xi)^{2n} = -(2n-1)\psi(\xi)^{2n-1}\frac{\partial\Lambda}{\partial\xi_\alpha}(\xi)\frac{\partial\psi}{\partial\bar{\xi}_\beta}(\xi) +$$

$$\psi(\xi)\left\{\frac{\partial^2\lambda}{\partial\xi_\alpha\partial\bar{\xi}_\beta}(\xi)\psi(\xi) + \frac{\partial\lambda}{\partial\xi_\alpha}(\xi)\frac{\partial\psi}{\partial\bar{\xi}_\beta}(\xi) - (2n-2)\left[\frac{\partial\lambda}{\partial\bar{\xi}_\beta}(\xi)\frac{\partial\psi}{\partial\xi_\alpha}(\xi) + \lambda(\xi)\frac{\partial^2\psi}{\partial\xi_\alpha\partial\bar{\xi}_\beta}(\xi)\right]\right\} .$$

Since $\lambda(\xi)$ is of class C^2 up to ∂D and $\psi(\xi_0) = 0$,

$$\lim_{\xi\to\xi_0}\frac{\partial^2\Lambda}{\partial\xi_\alpha\partial\bar{\xi}_\beta}(\xi)\psi(\xi)^{2n} = -(2n-1)\frac{\partial\psi}{\partial\bar{\xi}_\beta}(\xi_0)\lim_{\xi\to\xi_0}\left[\psi(\xi)^{2n-1}\frac{\partial\Lambda}{\partial\xi_\alpha}(\xi)\right] .$$

The right hand side of this equation equals

$$-(2n-1)(2n-2)\,\frac{\partial\psi}{\partial\bar{\xi}_\beta}(\xi_0)\,\frac{\partial\psi}{\partial\xi_\alpha}(\xi_0)\,\|\mathrm{Grad}\,\psi(\xi_0)\|^{2n-2}$$

by (2). This proves (3) . Equation (4) then follows from (1), (2), and (3). ∎

Note that for the unit ball $D = \{\xi \in \mathbb{C}^n\colon\ \psi(\xi) = \|\xi\|^2 - 1 < 0\}$,

$$\frac{\partial^2\log(-\Lambda)}{\partial\xi_\alpha\partial\bar{\xi}_\beta}(\xi)\,[\|\xi\|^2-1]^2 = (2n-2)[(1-\|\xi\|^2)\delta_{\alpha\beta} + \bar{\xi}_\alpha\xi_\beta]$$

$$= -(2n-2)\left[\psi(\xi)\delta_{\alpha\beta} - \frac{\partial\psi}{\partial\xi_\alpha}\frac{\partial\psi}{\partial\bar{\xi}_\beta}\right]$$

where $\delta_{\alpha\beta} = \begin{cases} 1, & \alpha = \beta \\ 0, & \alpha \neq \beta \end{cases}$. Thus, in this example,

$$\text{(6.11)} \qquad ds^2 \equiv \sum_{\alpha,\beta=1}^{n} \frac{\partial^2\log(-\Lambda)}{\partial\xi_\alpha\partial\bar{\xi}_\beta}(\xi)\,d\xi_\alpha \otimes d\bar{\xi}_\beta$$

is, up to a constant, the Bergman metric; (4) expresses the precise boundary behavior of the coefficients of the metric tensor. In Chapter 8 we will see that (6.11) defines a Kähler metric in case D is a bounded pseudoconvex domain with smooth boundary. For the moment, without the pseudoconvexity assumption, we can use (4) to show that curves which approach ∂D in a non-complex-tangential manner have infinite 'length' with respect to the 'metric' ds^2 . Precisely, we have the following result.

<u>Theorem 6.1</u>. Let $D \subset \mathbb{C}^n$ be a domain with smooth boundary. Let $\xi_0 \in \partial D$ and let $\gamma\colon\ t \to \xi(t)$, $0 \le t \le 1$, be a C^1 curve such that $\xi(t) \in D$ for $0 \le t < 1$, $\xi(1) = \xi_0$, and

for $0 \le t < 1$, $\xi(1) = \xi_0$, and

$$\sum_{\alpha=1}^{n} \frac{\partial\psi}{\partial\xi_\alpha}(\xi_0)\frac{d\xi_\alpha}{dt}(1) \neq 0 \tag{6.12}$$

$\left(\text{here, } \frac{d\xi_\alpha}{dt}(1) \equiv \lim_{t\to 1^-} \frac{d\xi_\alpha}{dt}(t)\right.$ which we assume exists, and ψ is a defining function for $\left.D\right)$.

Call $ds_\gamma^2(t) \equiv \sum_{\alpha,\beta=1}^{n} \frac{\partial^2 \log(-\Lambda)}{\partial\xi_\alpha\partial\bar{\xi}_\beta}(\xi(t))\frac{d\xi_\alpha}{dt}(t)\overline{\frac{d\xi_\beta}{dt}}(t)$ for $0 \le t < 1$.

Then

(a) there exists $0 < t_0 < 1$ such that $ds_\gamma^2(t) > 0$ for all $t_0 < t < 1$ and

(b) $\lim_{t\to 1^-} \int_{t_0}^{t} [ds_\gamma^2(t)]^{1/2}\, dt = +\infty$.

Proof. By Lemma 6.1 (4)

$$\begin{aligned}\lim_{t\to 1^-} ds_\gamma^2(t)\psi(\xi(t))^2 &= \lim_{t\to 1^-} \sum_{\alpha,\beta=1}^{n} \frac{\partial^2 \log(-\Lambda)}{\partial\xi_\alpha\partial\bar{\xi}_\beta}(\xi(t))\psi(\xi(t))^2\,\overline{\frac{d\xi_\beta}{dt}}(t)\frac{d\xi_\alpha}{dt}(t)\\ &= (2n-2)\left|\sum_{\alpha=1}^{n}\frac{\partial\psi}{\partial\xi_\alpha}(\xi_0)\frac{d\xi_\alpha}{dt}(1)\right|^2 > 0\end{aligned}$$

by assumption (6.12). Since γ is C^1, there exists $0 < t_0 < 1$ such that

$$ds_\gamma^2(t)\psi(\xi(t))^2 > (n-1)\left|\sum_{\alpha=1}^{n}\frac{\partial\psi}{\partial\xi_\alpha}(\xi(t))\frac{d\xi_\alpha}{dt}(t)\right|^2 > 0 \quad \text{for } t_0 < t < 1$$

which gives (a). It follows that, for such t,

$$ds_\gamma^2(t) > \frac{n-1}{4}\left|\frac{d\log|\psi(\xi(t))|}{dt}\right|^2$$

so that

$$\lim_{t\to1^-}\int_{t_0}^{t}[ds_\gamma^2(t)]^{1/2}dt > \frac{(n-1)^{1/2}}{2}\lim_{t\to1^-}\left|\log\frac{\psi(\xi(t))}{\psi(\xi(t_0))}\right| = +\infty$$

which is (b) . ∎

If (6.12) is not satisfied, then the theorem need not be true. Take, for example, the complement of the closed unit ball in $\mathbb{C}^2$, i.e.,

$$\tilde{D} \equiv \{\xi \in \mathbb{C}^2: \quad \psi(\xi) \equiv 1 - \|\xi\|^2 < 0\} .$$

Then the Robin function for $(\tilde{D},\xi)$ is given by

$$\tilde{\Lambda}(\xi) \equiv \frac{-1}{(\|\xi\|^2-1)^2} \quad \text{for} \quad \|\xi\| > 1 .$$

If we set $\xi_0 = (1,0) \in \partial\tilde{D}$ and γ: $t \to \xi(t) = (1,1-t)$, $0 \le t \le 1$, then it is easily verified that (6.12) does not hold. In this case,

$$\begin{aligned} d\tilde{s}_\gamma^{\,2}(t) &= \sum_{\alpha,\beta=1}^{2}\frac{\partial^2\log(-\tilde{\Lambda})}{\partial\xi_\alpha\partial\bar{\xi}_\beta}(\xi(t))\,\frac{d\xi_\alpha}{dt}(t)\,\overline{\frac{d\xi_\beta}{dt}}(t) \\ &= -\frac{\partial^2\log(\|\xi\|^2-1)}{\partial\xi_2\partial\bar{\xi}_2}(1,1-t) \equiv 0 \end{aligned}$$

for all $t \in (0,1)$. We can even construct an example of a bounded domain D and a curve γ in D approaching ∂D in a complex-tangential manner (so (6.12) fails) with $ds_\gamma^2(t)$ bounded near $t = 1$ (so (b) fails). Indeed, let

$$D = \{\xi \in \mathbb{C}^2: \quad 1 < \|\xi\| < 2\} \subset \tilde{D}$$

and let $\Lambda(\xi)$ be the Robin function for (D,ξ). Letting $\tilde{g}, g$ denote the Green function for $(\tilde{D},\xi)$, (D,ξ), the difference

$$u(\xi,z) \equiv \tilde{g}(\xi,z) - g(\xi,z)$$

is positive and harmonic as a function of $z \in D$ for each fixed ξ and equals 0 for $z \in \partial\tilde{D}$. Since u is symmetric in ξ and z, it follows that u has a real analytic extension to

$$\left\{ (\xi,z): \ \frac{1}{2} < \|\xi\| < 2 \ , \ \frac{1}{2} < \|z\| < 2 \right\}$$

In particular, $u(\xi,\xi) = \tilde{\Lambda}(\xi) - \Lambda(\xi)$ is real analytic in a neighborhood of $\|\xi\| = 1$ (precisely, we can define $\tilde{\Lambda}(\xi) \equiv u(\xi,\xi) + \Lambda(\xi)$ for $\frac{1}{2} < \|\xi\| \le 1$) so that

$$\begin{aligned}\Lambda(\xi) &= \tilde{\Lambda}(\xi) - u(\xi,\xi) \\ &= -\left[\frac{1 + u(\xi,\xi)(\|\xi\|^2-1)^2}{(\|\xi\| -1)^2}\right],\end{aligned}$$

and, with γ as before,

$$\begin{aligned}ds_\gamma^2(t) &= \left[\frac{\partial^2 \log[1+u(\xi,\xi)(\|\xi\|^2-1)^2]}{\partial\xi_2\partial\bar{\xi}_2}\right](1,1-t) + d\tilde{s}_\gamma^{\,2}(t) \\ &= \left[\frac{\partial^2 \log[1+u(\xi,\xi)(\|\xi\|^2-1)^2]}{\partial\xi_2\partial\bar{\xi}_2}\right](1,1-t)\end{aligned}$$

which is clearly bounded for t near 1.

We make a final remark concerning boundary behavior of the Robin function, Λ. In the case where ds^2 (defined in (6.11)) is a metric, the differential equation for a geodesic curve involves third order derivatives of Λ. Thus in order to study the boundary behavior of geodesic curves, the following result, which is an extension of Lemma 6.1, may be useful.

Lemma 6.2. Let D be a domain in $\mathbb{C}^n$ with smooth boundary. Let $\xi_0 \in \partial D$. Then for $\xi \in D$,

(6.13)

$$\lim_{\xi\to\xi_0} \frac{\partial^3\Lambda}{\partial\xi_\alpha\partial\bar{\xi}_\beta\partial\xi_\gamma}(\xi)\,\psi(\xi)^{2n+1}$$

$$= 2n(2n-1)(2n-2)\frac{\partial\psi}{\partial\xi_\alpha}(\xi_0)\frac{\partial\psi}{\partial\bar{\xi}_\beta}(\xi_0)\frac{\partial\psi}{\partial\xi_\gamma}(\xi_0)\,\|\mathrm{Grad}\,\psi(\xi_0)\|^{2n-2}.$$

Proof. Define

$$G_{\alpha\bar{\beta}\gamma}(\xi,z) \equiv \left(\frac{\partial G_{\alpha\bar{\beta}}}{\partial\xi_\gamma} + \frac{\partial G_{\alpha\bar{\beta}}}{\partial z_\gamma}\right)(\xi,z),\quad \alpha,\beta,\gamma = 1,\dots,n,$$

for $(\xi,z) \in D\times D$. As in the proof of Proposition 6.1, we can show that each $G_{\alpha\bar{\beta}\gamma}$ is real analytic and symmetric in $D\times D$, harmonic in z and in ξ, and satisfies

$$(6.14)\qquad \frac{\partial^3\Lambda}{\partial\xi_\alpha\partial\bar{\xi}_\beta\partial\xi_\gamma}(\xi) = G_{\alpha\bar{\beta}\gamma}(\xi,\xi) = 2\frac{\partial G_{\alpha\bar{\beta}}}{\partial z_\gamma}(\xi,\xi) = 2\frac{\partial G_{\alpha\bar{\beta}}}{\partial\xi_\gamma}(\xi,\xi).$$

Related to $G_{\alpha\bar{\beta}}$ (see (6.6)), the functions

(6.15)

$$g_{\alpha\bar{\beta}}(\xi,w) \equiv -(2n-1)\frac{\partial\psi}{\partial\bar{\xi}_\beta}g_\alpha + \psi\frac{\partial g_\alpha}{\partial\bar{\xi}_\beta} - \frac{\partial\psi}{\partial\bar{\xi}_\beta}\sum_{\beta=1}^{n}\left[w_\beta\frac{\partial g_\alpha}{\partial w_\beta} + \bar{w}_\beta\frac{\partial g_\alpha}{\partial\bar{w}_\beta}\right],$$

$$\alpha,\beta = 1,\dots,n,$$

are harmonic functions of $w \in D(\xi)$ for $\xi \in \bar{D}$, continuous for $(\xi, w) \in \mathcal{D}_{\bar{D}}$, and satisfy

$$g_{\alpha\bar{\beta}}(\xi, w) = G_{\alpha\bar{\beta}}(\xi, z)\ \psi(\xi)^{2n} \quad \text{for} \quad (\xi, w) \in \mathcal{D}_D \tag{6.16}$$

This can be seen by differentiating (6.9) with respect to $\bar{w}_\beta$ and $\bar{\xi}_\beta$.

To prove (6.13), we compute $\frac{\partial G_{\alpha\bar{\beta}}}{\partial z_\gamma}(\xi, \xi)$ and use (6.14).
Differentiate (6.16) with respect to z_γ and set $z = \xi$. Using (6.15), we obtain

$$\frac{\partial G_{\alpha\bar{\beta}}}{\partial z_\gamma}(\xi, \xi) = \frac{-1}{\psi(\xi)^{2n+1}} \frac{\partial g_{\alpha\bar{\beta}}}{\partial w_\gamma}(\xi, 0)$$

$$= \frac{-1}{\psi(\xi)^{2n+1}} \left\{ (-2n) \frac{\partial \psi}{\partial \bar{\xi}_\beta}(\xi) \frac{\partial g_\alpha}{\partial w_\gamma}(\xi, 0) + \psi(\xi) \frac{\partial^2 g_\alpha}{\partial w_\gamma \partial \bar{\xi}_\beta}(\xi, 0) \right\}.$$

Since $\frac{\partial g_\alpha}{\partial w_\gamma}(\xi, 0)$ and thus also $\frac{\partial^2 g_\alpha}{\partial w_\gamma \partial \bar{\xi}_\beta}(\xi, 0)$ are continuous at $\xi_0 \in \partial D$,

$$\lim_{\xi \to \xi_0} \frac{\partial G_{\alpha\bar{\beta}}}{\partial z_\gamma}(\xi, \xi)\ \psi(\xi)^{2n+1} = 2n \frac{\partial \psi}{\partial \bar{\xi}_\beta}(\xi_0) \frac{\partial g_\alpha}{\partial w_\gamma}(\xi_0, 0). \tag{6.17}$$

It remains to compute $\frac{\partial g_\alpha}{\partial w_\gamma}(\xi_0, 0)$. Differentiate (6.7) with respect to w_γ and set $(\xi, w) = (\xi_0, 0)$ to obtain

$$\frac{\partial g_\alpha}{\partial w_\gamma}(\xi_0, 0) = -(n-1) \frac{\partial \psi}{\partial \xi_\alpha}(\xi_0) \frac{\partial}{\partial w_\gamma} \left[g_0(\xi_0, w) + \overline{g_0(\xi_0, w)} \right] \Big|_{w=0}. \tag{6.18}$$

Using the explicit formula (6.10) for $g_0(\xi_0, w)$ we compute that

$$(6.19) \quad \frac{\partial}{\partial w_\gamma} [g_0(\xi_0, w) + \overline{g_0(\xi_0, w)}]\Big|_{w=0} = -(2n-1) \frac{\partial \psi}{\partial \xi_\gamma} (\xi_0) \, \|\mathrm{Grad}\, \psi(\xi_0)\|^{2n-2}$$

Thus (6.14) , together with (6.17)-(6.19) yield

$$\lim_{\xi \to \xi_0} \frac{\partial^3 \Lambda}{\partial \xi_\alpha \partial \bar{\xi}_\beta \partial \xi_\gamma} (\xi) \, \psi(\xi)^{2n+1} = 2 \lim_{\xi \to \xi_0} \frac{\partial G_{\alpha\bar{\beta}}}{\partial z_\gamma} (\xi, \xi) \, \psi(\xi)^{2n+1}$$

$$= 4n(n-1)(2n-1) \frac{\partial \psi}{\partial \xi_\alpha} (\xi_0) \frac{\partial \psi}{\partial \bar{\xi}_\beta} (\xi_0) \frac{\partial \psi}{\partial \xi_\gamma} (\xi_0) \|\mathrm{Grad}\, \psi(\xi_0)\|^{2n-2}$$

which is (6.13). ■

7. STRICT PLURISUBHARMONICITY OF $-\Lambda(\xi)$, $\log(-\Lambda(\xi))$

In this section, we use the fundamental formula (Corollary 2.1) from Chapter 2 to derive a formula for the Levi forms of the functions $-\Lambda(\xi)$ and $\log(-\Lambda(\xi))$ associated with a smoothly bounded domain $D \subset \mathbb{C}^n$. Then using a differential inequality characterizing pseudoconvex boundary points for such domains (Proposition 7.1), we give a new proof of Theorem 8.1 [Y] which states that if D is pseudoconvex and bounded with smooth boundary, then $-\Lambda(\xi)$ and $\log(-\Lambda(\xi))$ are strictly plurisubharmonic functions in D.

Let $D \subset \mathbb{C}^n$ be a domain with smooth boundary and let $(\tilde{D}, \psi)$ be a double defining D. As usual, we let $G(\xi, z)$ and $\Lambda(\xi)$ denote the Green function and Robin constant for (D, ξ). Fix a non-zero vector $a \in \mathbb{C}^n$ and a point $\xi_0 \in D$. To get a formula for the Levi form

$$\sum_{\alpha,\beta=1}^{n} \frac{\partial^2(-\Lambda)}{\partial\xi_\alpha \partial\bar{\xi}_\beta}(\xi_0) a_\alpha \bar{a}_\beta ,$$

we follow the procedure in [Y]. Let $B = \{t \in \mathbb{C}: |t| < \rho\}$ be a disc such that $\xi_0 + at \in D$ for all $t \in B$. The transformation

$$T_1(t, z) = (t, w) \equiv (t, z-at)$$

transforms the product domain $B \times D$ $(B \times \tilde{D})$ onto the domain $\mathcal{D}_1 \equiv T_1(B \times D)$ $(\tilde{\mathcal{D}}_1 \equiv T_1(B \times \tilde{D}))$ in a biholomorphic fashion. If we set $\psi_1(t, w) = \psi(w+at)$ for $(t, w) \in \tilde{\mathcal{D}}_1$, then clearly the double $(\tilde{\mathcal{D}}_1, \psi_1)$ defines the smooth variation

$$\mathcal{D}_1: \quad t \to D_1(t)$$

on B. Note that $B \times \{\xi_0\} \subset \mathcal{D}$, so that there exist a Green function $g_1(t, w)$ and Robin constant $\lambda_1(t)$ for $(D_1(t), \xi_0)$. Furthermore,

$$\sum_{\alpha,\beta=1}^{n} \frac{\partial^2(-\Lambda)}{\partial\xi_\alpha\partial\bar{\xi}_\beta}(\xi_0)a_\alpha\bar{a}_\beta = \frac{\partial^2(-\lambda_1)}{\partial t\partial\bar{t}}(0) ,$$

and, by Corollary 2.1,

$$(7.1) \qquad \frac{\partial^2(-\lambda_1)}{\partial t\partial\bar{t}}(0) = \frac{1}{(n-1)w_{2n}}\int_{\partial D_1(0)} k_2(0,w)\|\mathrm{Grad}_{(w)}g_1(0,w)\|^2 dS_w$$

$$+ \frac{4}{(n-1)w_{2n}}\iint_{\partial D_1(0)} \left[\sum_{\alpha=1}^{n}\left|\frac{\partial^2 g_1(0,w)}{\partial\bar{t}\partial w_\alpha}\right|^2\right] dV_w$$

$$\equiv I(\xi_o,a) + J(\xi_0,a)$$

where $D_1(0) = D$,

$$k_2(0,w) = \mathcal{L}\psi_1(0,w)/\|\mathrm{Grad}_{(w)}\psi_1(0,w)\|^3 \text{ , and}$$

$$\mathcal{L}\psi_1(0,w) = \left[\left|\frac{\partial\psi_1}{\partial t}\right|^2 \Delta_{(w)}\psi_1 - 2\mathrm{Re}\left\{\sum_{\alpha=1}^{n}\frac{\partial\psi_1}{\partial\bar{t}}\frac{\partial^2\psi_1}{\partial t\partial\bar{w}_\alpha}\frac{\partial\psi_1}{\partial w_\alpha}\right\}\right.$$

$$\left.+ \frac{\partial^2\psi_1}{\partial t\partial\bar{t}}\|\mathrm{Grad}_{(w)}\psi_1\|^2\right]\Bigg|_{(0,w)} .$$

We want to rewrite $I(\xi_0,a)$ and $J(\xi_0,a)$ in terms of the original coordinates z and ξ using the functions $G(\xi,z)$, $\psi(z)$ and their derivatives. Since

$$(7.2) \qquad g_1(t,w) = G(\xi_0+at,z) \text{ and } \lambda_1(t) = \Lambda(\xi_0+at) ,$$

we compute, recalling the definition of $G_{\bar{\beta}}$ in (6.5) and using the fact that $z = w$ if $t = 0$,

$$\frac{\partial g_1}{\partial t}(0,w) = \sum_{\beta=1}^{n} a_\beta G_\beta(\xi_0,z) \quad \text{and}$$

$$\frac{\partial^2 g_1}{\partial t \partial \bar{w}_\alpha}(0,w) = \sum_{\beta=1}^{n} a_\beta \frac{\partial}{\partial \bar{z}_\alpha} G_\beta(\xi_0,z) .$$

Hence

$$J(\xi_0,a) = \frac{4}{(n-1)w_{2n}} \iint_D \sum_{\alpha=1}^{n} \left| \sum_{\beta=1}^{n} a_\beta \frac{\partial}{\partial \bar{z}_\alpha} G_\beta(\xi_0,z) \right|^2 dV_z \tag{7.3}$$

(equation (8.5), [Y]) . To rewrite $I(\xi_0,a)$, since $\psi_1(t,w) = \psi(w+at)$ direct computation yields

$$k_2(0,w) = \frac{1}{\|\text{Grad } \psi\|^3} \left\{ \left| \sum_{\alpha=1}^{n} \frac{\partial \psi}{\partial z_\alpha} a_\alpha \right|^2 \Delta\psi - 2\text{Re}\left[\left(\sum_{\alpha=1}^{n} \frac{\partial \psi}{\partial z_\alpha} a_\alpha \right) \left(\sum_{i,\beta=1}^{n} \frac{\partial \psi}{\partial \bar{z}_i} \frac{\partial^2 \psi}{\partial z_i \partial \bar{z}_\beta} \bar{a}_\beta \right) \right] \right.$$

$$\left. + \left(\sum_{\alpha,\beta=1}^{n} \frac{\partial^2 \psi}{\partial z_\alpha \partial \bar{z}_\beta} a_\alpha \bar{a}_\beta \right) \|\text{Grad } \psi\|^2 \right\} \tag{7.4}$$

$$\equiv K_2(z,a) .$$

The notation suggests that $K_2(z,a)$ is independent of the defining function ψ chosen; indeed, this may be verified by direct computation. Thus we have

$$I(\xi_0,a) = \frac{1}{(n-1)w_{2n}} \int_{\partial D} K_2(z,a) \|\text{Grad}_{(z)} G(\xi_0,z)\|^2 \, dS_z \tag{7.5}$$

so that

$$\sum_{\alpha,\beta=1}^{n} \frac{\partial^2(-\Lambda)}{\partial \xi_\alpha \partial \bar{\xi}_\beta}(\xi_0) a_\alpha \bar{a}_\beta = \frac{1}{(n-1)w_{2n}} \int_{\partial D} K_2(z,a) \|\text{Grad}_{(z)} G(\xi_0,z)\|^2 \, dS_z$$

$$+ \frac{4}{(n-1)w_{2n}} \iint_D \sum_{\alpha=1}^{n} \left| \sum_{\beta=1}^{n} a_\beta \frac{\partial}{\partial \bar{z}_\alpha} G_\beta(\xi_o,z) \right|^2 dV_z \tag{7.6}$$

To get a formula for

$$\sum_{\alpha,\beta=1}^{n} \frac{\partial^2 \log(-\Lambda)}{\partial\xi_\alpha \partial\bar{\xi}_\beta} (\xi_0) a_\alpha \bar{a}_\beta$$

as in (8.11) [Y], we consider the Hartogs transformation

$$T_2(t,w) = (t,W) \equiv \left(t,\ e^{\frac{\phi(t)}{2n-2}} (w-\xi_0)\right)$$

where

$$\phi(t) = c_0 + c_1 t\ ,\quad c_0 = \log(-\lambda_1(0)) = \log(-\Lambda(\xi_0))\ ,$$

and

$$c_1 = 2\,\frac{\partial \log(-\lambda_1)}{\partial t}(0) = 2\sum_{\alpha=1}^{n} \frac{\partial \log(-\lambda_1)}{\partial \xi_\alpha}(\xi_0) a_\alpha\ .$$

Setting $\mathcal{D}_2 = T_2(\mathcal{D}_1)$, $\tilde{\mathcal{D}}_2 = T_2(\tilde{\mathcal{D}}_1)$ and $\psi_2(t,W) = \psi_1(t,w)$, the double $(\tilde{\mathcal{D}}_2, \psi_2)$ defines the smooth variation

$$\mathcal{D}_2 :\quad t \to D_2(t)$$

on B. Since $B \times \{0\} \subset \mathcal{D}_2$, there exist a Green function $g_2(t,W)$ and Robin constant $\lambda_2(t)$ for each $(D_2(t),0)$. This time we have

$$\sum_{\alpha,\beta=1}^{n} \frac{\partial^2 \log(-\Lambda)}{\partial\xi_\alpha \partial\bar{\xi}_\beta} (\xi_0) a_\alpha \bar{a}_\beta = \frac{\partial^2(-\lambda_2)}{\partial t \partial\bar{t}}(0)$$

and, by Corollary 2.1,

$$\frac{\partial^2(-\lambda_2)}{\partial t\partial\bar{t}}(0) = \frac{1}{(n-1)w_{2n}} \int_{\partial D_2(0)} k_2(0,W) \|\mathrm{Grad}_{(W)} g_2(0,W)\|^2 dS_W$$

(7.7)
$$+ \frac{4}{(n-1)w_{2n}} \iint_{D_2(0)} \left[\sum_{\alpha=1}^{n} \left| \frac{\partial^2 g_2}{\partial\bar{t}\partial W_\alpha}(0,W)\right|^2 \right] dV_W \equiv \tilde{I}(\xi_0, a) + \tilde{J}(\xi_0, a)\ .$$

Using the relations

$$(7.8)\qquad \begin{aligned} g_2(t,W) &= e^{-\mathrm{Re}\phi(t)} G(\xi_0+at, We^{-\frac{\phi(t)}{2n-2}} + \xi_0 + at) \quad \text{and} \\ \lambda_2(t) &= e^{-\mathrm{Re}\phi(t)}\lambda_1(t) = e^{-\mathrm{Re}\phi(t)}\, \Lambda(\xi_0+at) , \end{aligned}$$

the volume integral $\tilde{J}(\xi_0, a)$ may be written as

$$(7.9)\qquad \tilde{J}(\xi_0,a) = \frac{4}{(n-1)w_{2n}(-\Lambda(\xi_0))} \iint_D \left[\sum_{\beta=1}^{n} \left| \frac{\partial}{\partial \bar{z}_\beta} H(a,\xi_0,z) \right|^2 \right] dV_z$$

where

$$(7.10)\qquad H(a,\xi_0,z) = \frac{-c_1}{2} G_0(\xi,z) + \sum_{\beta=1}^{n} a_\beta G_\beta(\xi,z)$$

and

$$(7.11)\qquad G_0(\xi,z) = G(\xi,z) + \frac{1}{n-1} \sum_{\beta=1}^{n} (z_\beta - \xi_\beta) \frac{\partial G}{\partial z_\beta}(\xi,z)$$

(cf. (8.11) [Y]). To rewrite $\tilde{I}(\xi_0, a)$, note that

$$\psi_2(t,W) = \psi_2(T_2 \circ T_1(t,z)) = \psi\left(W\, e^{\frac{-c_0-c_1 t}{2n-2}} + \xi_0 + at\right) .$$

Thus

$$L_{(t,W_i)}\psi_2(0,W) \equiv \left[\frac{\partial^2\psi_2}{\partial t\partial\bar{t}} \left|\frac{\partial\psi_2}{\partial W_i}\right|^2 - 2\mathrm{Re}\left\{ \frac{\partial\psi_2}{\partial t} \frac{\partial^2\psi_2}{\partial\bar{t}\partial W_i} \frac{\partial\psi_2}{\partial\bar{W}_i} \right\} + \left|\frac{\partial\psi_2}{\partial t}\right|^2 \frac{\partial^2\psi_2}{\partial W_i\partial\bar{W}_i} \right] \Bigg|_{(0,W)}$$

$$= \left\{ \sum_{\alpha,\beta=1}^{n} \frac{\partial^2\psi}{\partial z_\alpha \partial\bar{z}_\beta} \left(\frac{-c_1}{2n-2} e^{\frac{-c_0}{2n-2}} W_\alpha + a_\alpha \right) \overline{\left(\frac{-c_1}{2n-2} e^{\frac{-c_0}{2n-2}} W_\beta + a_\beta \right)} \right\} \left| \frac{\partial\psi}{\partial z_i} e^{\frac{-c_0}{2n-2}} \right|^2$$

$$- 2\mathrm{Re}\left[\left\{\sum_{\alpha=1}^{n} \frac{\partial\psi}{\partial z_\alpha}\left(\frac{-c_1}{2n-2}\, e^{\frac{-c_0}{2n-2}} W_\alpha + a_\alpha\right)\right\}\left\{\sum_{\beta=1}^{n} \frac{\partial^2\psi}{\partial\bar{z}_\beta \partial z_i}\overline{\left(\frac{-c_1}{2n-2}\, e^{\frac{-c_0}{2n-2}} W_\beta + a_\beta\right) e^{\frac{-c_0}{2n-2}}}\right\} \cdot \overline{\left\{\frac{\partial\psi}{\partial z_i}\, e^{\frac{-c_0}{2n-2}}\right\}}\right.$$

$$+ \left|\sum_{\alpha=1}^{n} \frac{\partial\psi}{\partial z_\alpha}\left(\frac{-c_1}{2n-2}\, e^{\frac{-c_0}{2n-2}} W_\alpha + a_\alpha\right)\right|^2 \left(\frac{\partial^2\psi}{\partial z_i \partial\bar{z}_i}\left|e^{\frac{-c_0}{2n-2}}\right|^2\right) .$$

Note that at $t = 0$, $z - \xi_0 = e^{\frac{-c_0}{2n-2}} W$. If we define

$$(7.12) \qquad \mathcal{O}(a, \xi_0, z) \equiv \frac{-c_1}{2n-2}(z-\xi_0) + a ,$$

then

$$L_{(t, W_i)}\psi_2(0, W) = e^{\frac{-c_0}{n-1}}\left\{\left[\sum_{\alpha,\beta=1}^{n} \frac{\partial^2\psi}{\partial z_\alpha \partial\bar{z}_\beta}\mathcal{O}_\alpha\bar{\mathcal{O}}_\beta\right]\left|\frac{\partial\psi}{\partial z_i}\right|^2\right.$$

$$\left. - 2\,\mathrm{Re}\left[\sum_{\alpha,\beta=1}^{n}\left(\frac{\partial\psi}{\partial z_\alpha}\mathcal{O}_\alpha\right)\frac{\partial^2\psi}{\partial\bar{z}_\beta \partial z_i}\bar{\mathcal{O}}_\beta\right)\frac{\partial\psi}{\partial\bar{z}_i}\right] + \left|\sum_{\alpha=1}^{n} \frac{\partial\psi}{\partial z_\alpha}\mathcal{O}_\alpha\right|^2 \frac{\partial^2\psi}{\partial z_i \partial\bar{z}_i}\right\}$$

so that

$$k_2(0, W) = \sum_{i=1}^{n} L_{(t, W_i)}\psi_2(0, W)/\|\mathrm{Grad}_{(W)}\psi_2(0, W)\|^3$$

$$= \frac{e^{\frac{c_0}{2n-2}}}{\|\mathrm{Grad}\psi\|^3}\left\{\sum_{\alpha,\beta=1}^{n} \frac{\partial^2\psi}{\partial z_\alpha \partial\bar{z}_\beta}\mathcal{O}_\alpha\bar{\mathcal{O}}_\beta\|\mathrm{Grad}\ \psi\|^2 - 2\mathrm{Re}\left[\sum_{\alpha,\beta,i=1}^{n} \frac{\partial\psi}{\partial z_\alpha}\frac{\partial^2\psi}{\partial\bar{z}_\beta \partial z_i}\frac{\partial\psi}{\partial z_i}\mathcal{O}_\alpha\bar{\mathcal{O}}_\beta\right]\right.$$

$$+ \left|\sum_{\alpha=1}^{n} \frac{\partial\psi}{\partial z_\alpha} \mathcal{O}_\alpha\right|^2 \Delta_{(z)}\psi\Bigg\} = e^{\frac{c_0}{2n-2}} K_2(z,\mathcal{O})$$

from (7.4) for $W \in \partial D_2(0)$, where $z = \xi_0 + e^{\frac{-c_0}{2n-2}} W \in \partial D$. By (7.8)

$$g_2(0,W) = e^{-c_0} G\left(\xi_0, W e^{\frac{-c_0}{2n-2}} + \xi_0\right) \quad \text{so that}$$

$$\mathrm{Grad}_{(W)} g_2(0,W) = e^{-(\frac{2n-1}{2n-2})c_0} \mathrm{Grad}_{(z)} G(\xi_0, z) .$$

Thus $dS_W = e^{(\frac{2n-1}{2n-2})c_0} dS_z$, and using the fact that $e^{-c_0} = \frac{1}{-\Lambda(\xi_0)}$, we have

(7.13) $$\tilde{I}(\xi_0, a) = \frac{1}{(n-1)w_{2n}(-\Lambda(\xi_0))} \int_{\partial D} K_2(z,\mathcal{O}) \|\mathrm{Grad}_{(z)} G(\xi_0, z)\|^2 \, dS_z .$$

Hence

(7.14)

$$\sum_{\alpha,\beta=1}^{n} \frac{\partial^2 \log(-\Lambda)}{\partial\xi_\alpha \partial\bar{\xi}_\beta} (\xi_0) a_\alpha \bar{a}_\beta = \frac{1}{(n-1)w_{2n}(-\Lambda(\xi_0))} \int_{\partial D} K_2(z,\mathcal{O}) \|\mathrm{Grad}_{(z)} G(\xi_0, z)\|^2 \, dS_z$$

$$+ \frac{4}{(n-1)w_{2n}(-\Lambda(\xi_0))} \iint_D \left[\sum_{\beta=1}^{n} \left|\frac{\partial}{\partial \bar{z}_\beta} H(a, \xi_0, z)\right|^2 \right] dV_z .$$

We now analyze the quantity K_2 defined in (7.4) which occurs in both $I(\xi_0, a)$ and $\tilde{I}(\xi_0, a)$. In general, we can take D to be any domain in $\mathbb{C}^n$; the quantity $K_2(z_0, a)$ will detect pseudoconvexity at a nonsingular boundary point $z_0 \in \partial D$; i.e., at a point $z_0 \in \partial D$ such that there exist a neighborhood V of z_0 and a C^2 function $\psi = \psi(z)$ in V such that $D \cap V = \{z: \psi(z) < 0\}$, $\partial D \cap V = \{z: \psi(z) = 0\}$, and $\mathrm{Grad}\, \psi(z) \neq 0$ for

$z \in \partial D \cap V$. The classical definition of pseudoconvexity at z_0 is that the Levi form

$$\sum_{\alpha,\beta=1}^{n} \frac{\partial^2\psi}{\partial z_\alpha \partial \bar{z}_\beta}(z_0)\, a_\alpha \bar{a}_\beta$$

be nonnegative for all nonzero vectors a in $\mathbb{C}^n$ satisfying $\sum_{\alpha=1}^{n} \frac{\partial\psi}{\partial z_\alpha}(z_0)a_\alpha = 0$ (strictly positive for strict pseudoconvexity). Define $K_2\psi(z_0, a)$ using (7.4), i.e., for a nonsingular boundary point $z_0 \in \partial D$ and a nonzero vector a,

(7.15)

$$K_2\psi(z_0,a) \equiv \frac{1}{\|\mathrm{Grad}\ \psi\|^3}\left\{\left|\sum_{\alpha=1}^{n} \frac{\partial\psi}{\partial z_\alpha} a_\alpha\right|^2 \Delta\psi \right.$$

$$- 2\,\mathrm{Re}\left[\left(\sum_{\alpha=1}^{n} \frac{\partial\psi}{\partial z_\alpha} a_\alpha\right)\left(\sum_{i,\beta=1}^{n} \frac{\partial\psi}{\partial \bar{z}_i} \frac{\partial^2\psi}{\partial z_i \partial \bar{z}_\beta} \bar{a}_\beta\right)\right]$$

$$\left. + \left(\sum_{\alpha,\beta=1}^{n} \frac{\partial^2\psi}{\partial z_\alpha \partial \bar{z}_\beta} a_\alpha \bar{a}_\beta\right)\|\mathrm{Grad}\ \psi\|^2 \right\}\Bigg|_{z=z_0}.$$

Note the following properties of $K_2\psi$:

(1) $K_2\psi(z_0, ta) = |t|^2 K_2\psi(z_0,a)$ for $t \in \mathbb{C}$.

(2) $K_2\psi(z_0,a)$ is invariant under unitary transformations; i.e., if $\tilde{z} = Az$ is a unitary transformation of $\mathbb{C}^n$, $\tilde{D} = A(D)$, $\tilde{\psi}(\tilde{z}) = \psi(z)$, and $\tilde{a} = A(a)$, then $K_2\tilde{\psi}(\tilde{z},\tilde{a}) = K_2\psi(z,a)$.

(3) $K_2\psi(z_0,a)$ is independent of the defining function ψ; i.e., if $\phi = f\psi$ where $f > 0$ on V, then $K_2\phi(z_0,a) = K_2\psi(z_0,a)$. This last

property justifies the notation used in (7.4); i.e., we can write $K_2\psi(z_0, a) = K_2(z_0, a)$.

Proposition 7.1. Let $z_0 \in \partial D$ be nonsingular. Then D is pseudoconvex (strictly pseudoconvex) at z_0 if and only if $K_2(z_0, a) \geq 0$ (>0) for all $a \in \mathbb{C}^n - \{0\}$.

Proof. By the above properties (1) - (3), we may assume $\text{Grad } \psi(z_0) = (0, 0, \ldots, 1)$. Then

(7.16)

$$K_2(z_0, a) = \left(\frac{\partial^2 \psi}{\partial z_1 \partial \bar{z}_1} + \ldots + \frac{\partial^2 \psi}{\partial z_{n-1} \partial \bar{z}_{n-1}} \right)(z_0)|a_n|^2 + \sum_{\alpha, \beta=1}^{n-1} \frac{\partial^2 \psi}{\partial z_\alpha \partial \bar{z}_\beta}(z_0) a_\alpha \bar{a}_\beta .$$

First assume that D is pseudoconvex (strictly pseudoconvex) at z_0 . Then the second term on the right-hand side of (7.16) is ≥ 0 (>0) for $a' = (a_1, \ldots, a_{n-1}) \in \mathbb{C}^{n-1} - \{0\}$. In particular

$$\left(\frac{\partial^2 \psi}{\partial z_1 \partial \bar{z}_1} + \ldots + \frac{\partial^2 \psi}{\partial z_{n-1} \partial \bar{z}_{n-1}} \right)(z_0) \geq 0 (>0)$$

so that $K_2(z_0, a) \geq 0$ (>0) if $a = (a_1, \ldots a_n) \in \mathbb{C}^n - \{0\}$. The converse follows from setting $a_n = 0$. ∎

We now see the value of the calculations carried out in the beginning of this chapter to get (7.6) and (7.14); together with Proposition 7.1, we get a simple proof of Theorem 8.1 [Y] . Note that we did not need to assume the domain D was bounded for the previous work in this chapter.

Corollary 7.1. Let D be a bounded pseudoconvex domain in $\mathbb{C}^n$ with smooth boundary. Then $-\Lambda(\xi)$ and $\log(-\Lambda(\xi))$ are strictly plurisubharmonic in D .

Proof. Fix $\xi_0 \in D$ and $a \in \mathbb{C}^n - \{0\}$. By (7.6),

$$\sum_{\alpha,\beta=1}^{n} \frac{\partial^2(-\Lambda)}{\partial\xi_\alpha \partial\bar{\xi}_\beta}(\xi_0)a_\alpha \bar{a}_\beta \geq \frac{1}{(n-1)w_{2n}} \int_{\partial D} K_2(z,a)\|\mathrm{Grad}_{(z)}G(\xi_0,z)\|^2 dS_z .$$

By the proposition, $K_2(z,a) \geq 0$ for all $z \in \partial D$ so that $-\Lambda$ is plurisubharmonic. Furthermore, since D is bounded and ∂D is smooth, there exists a strictly pseudoconvex boundary point $z_0 \in \partial D$. Thus, again by the proposition, $K_2(z_0,a) > 0$, and this inequality remains valid for z near z_0 in ∂D. Hence $-\Lambda$ is strictly plurisubharmonic.

Using (7.14),

$$\sum_{\alpha,\beta=1}^{n} \frac{\partial^2 \log(-\Lambda)}{\partial\xi_\alpha \partial\bar{\xi}_\beta}(\xi_0)a_\alpha \bar{a}_\beta \geq \frac{1}{(n-1)w_{2n}(-\Lambda(\xi_0))} \int_{\partial D} K_2(z,\mathcal{O})\|\mathrm{Grad}_{(z)}G(\xi_0,z)\|^2 dS_z .$$

Again using the proposition, $K_2(z,\mathcal{O}) \geq 0$ for $z \in \partial D$ which gives the plurisubharmonicity of $\log(-\Lambda)$. Moreover, we can find a strictly pseudoconvex boundary point $z_0 \in \partial D$ such that

$$\mathcal{O} = \frac{-c_1}{2n-2}(z_0-\xi_0) + a \neq 0 .$$

Hence $K_2(z,\mathcal{O}) > 0$ for z near z_0 in ∂D so that $\log(-\Lambda)$ is strictly plurisubharmonic. ∎

8. THE ROBIN FUNCTION AND THE BERGMAN KERNEL

Note that from equation (7.6) in the last chapter, if D is pseudoconvex we have the inequality

$$(8.1) \quad \sum_{\alpha,\beta=1}^{n} \frac{\partial^2(-\Lambda)}{\partial\xi_\alpha\partial\bar{\xi}_\beta}(\xi)\, a_\alpha \bar{a}_\beta \geq \frac{4}{(n-1)w_{2n}} \iint_D \sum_{\alpha=1}^{n} \left| \sum_{\beta=1}^{n} a_\beta \frac{\partial G_\beta}{\partial \bar{z}_\alpha}(\xi, z) \right|^2 dV_z$$

for $\xi \in D$ and $a \in \mathbb{C}^n$. We now use (8.1) to relate the Bergman kernel function $K(\xi,z)$ for D with the Robin function $\Lambda(\xi)$.

We briefly recall the definition of $K(\xi,z)$. Let $D \subset \mathbb{C}^n$ be a domain with smooth boundary. We write

$$(f,g)_D \equiv \iint_D f(z)\,\overline{g(z)}\, dV_z \quad \text{and} \quad \|f\|_D^2 \equiv (f,f)_D$$

for $f,g \in L^2(D) \equiv \{f\colon\ D \to \mathbb{C} \mid \|f\|_D^2 < +\infty\}$. Given $\xi \in D$, there exists a unique holomorphic function $K(\xi,z)$ for $z \in D$ satisfying

$$f(\xi) = (f, K(\xi,\cdot))_D$$

for each holomorphic $f \in L^2(D)$. The function $K(\xi,z)$ defined in $D \times D$ is called the <u>Bergman Kernel Function</u> for D. In the case $n = 1$, using the Robin Function $\Lambda(\xi)$ associated with the logarithmic potential in $\mathbb{C}$, Suita [S] showed that

$$\frac{\partial^2(-\Lambda)}{\partial\xi\partial\bar{\xi}}(\xi) = \frac{\pi}{2} K(\xi,\xi)$$

for $\xi \in D$. In $\mathbb{C}^n$, $n \geq 2$, we have the following result.

<u>Theorem 8.1</u>. Let $D \subset \mathbb{C}^n$ be pseudoconvex with smooth boundary. Then for $\xi \in D$,

$$(8.2) \qquad \sum_{\alpha=1}^{n} \frac{\partial^2(-\Lambda)}{\partial\xi_\alpha \partial\bar{\xi}_\alpha}(\xi) \geq \frac{(n-1)w_{2n}}{n} K(\xi,\xi) .$$

Remark 1. If $n = 1$, every domain is pseudoconvex and we have an exact relationship between $K(\xi,\xi)$ and $\Lambda(\xi)$ (Suita's result), while we can only prove an inequality if $n > 1$.

2. To illustrate the theorem, recall that for the unit ball $D = \{\xi \in \mathbb{C}^n: \ \|\xi\| < 1\}$ we have

$$\Lambda(\xi) = \frac{-1}{(1-\|\xi\|^2)^{2n-2}} \quad \text{and} \quad K(\xi,\xi) = \frac{n!}{\pi^n} \frac{1}{(1-\|\xi\|^2)^{n+1}} .$$

Since $w_{2n} = \dfrac{2\pi^n}{(n-1)!}$, we have

$$\sum_{\alpha=1}^{n} \frac{\partial^2(-\Lambda)}{\partial\xi_\alpha \partial\bar{\xi}_\alpha}(\xi) = \frac{2(n-1)}{(1-\|\xi\|^2)^{2n}}[(n-1)\|\xi\|^2+n] > \frac{2(n-1)}{(1-\|\xi\|^2)^{n+1}} = \frac{(n-1)w_{2n}}{n} K(\xi,\xi) .$$

As this example indicates, the inequality (8.2) is far from being sharp - this will also be clear from the proof of the theorem. We begin with a lemma.

Lemma 8.1. Let $D \subset \mathbb{C}^n$ be a domain with smooth boundary. Define $H(\xi,z) \equiv \sum_{\alpha=1}^{n} \dfrac{\partial^2 G}{\partial\bar{\xi}_\alpha \partial z_\alpha}(\xi,z)$ for $(\xi,z) \in D \times D$. Then

(1) $H(\xi,z)$ is real analytic in $D \times D$ and is harmonic in ξ and in z.

(2) $H(\xi,\xi) = \dfrac{1}{2} \sum_{\alpha=1}^{n} \dfrac{\partial^2(-\Lambda)}{\partial\xi_\alpha \partial\bar{\xi}_\alpha}(\xi)$ for $\xi \in D$.

(3) $H(\xi,z) = \overline{H(z,\xi)}$.

(4) For any $f \in C^1(\bar{D})$ which is harmonic in D,

$$f(\xi) = \frac{-2}{(n-1)w_{2n}} \iint_D [f(z)\overline{H(\xi,z)} + \sum_{\alpha=1}^{n} \frac{\partial f}{\partial \bar{z}_\alpha}(z)\, G_\alpha(\xi,z)]dV_z$$

for all $\xi \in D$.

Proof. From the definition of $G_\alpha(\xi,z)$ in (6.5) and the harmonicity of $G(\xi,z)$ in ξ, it follows that

$$\sum_{\alpha=1}^{n} \frac{\partial G_\alpha}{\partial \bar{\xi}_\alpha}(\xi,z) = \sum_{\alpha=1}^{n} \frac{\partial^2 G}{\partial \bar{\xi}_\alpha \partial z_\alpha}(\xi,z) = H(\xi,z)\,.$$

Properties (1) and (2) then follow from Proposition 6.1. If we differentiate the symmetry formula $G_\alpha(\xi,z) = G_\alpha(z,\xi)$, we obtain

$$\frac{\partial G_\alpha}{\partial \bar{\xi}_\alpha}(\xi,z) = \frac{\partial G_\alpha}{\partial \bar{z}_\alpha}(z,\xi) = \frac{\partial^2 G}{\partial z_\alpha \partial \bar{z}_\alpha}(z,\xi) + \frac{\partial^2 G}{\partial \xi_\alpha \partial \bar{z}_\alpha}(z,\xi)$$

so that, using harmonicity of $G(\xi,z)$ in z,

$$H(\xi,z) = \sum_{\alpha=1}^{n} \frac{\partial G_\alpha}{\partial \bar{\xi}_\alpha}(\xi,z) = \sum_{\alpha=1}^{n} \frac{\partial^2 G}{\partial \xi_\alpha \partial \bar{z}_\alpha}(z,\xi) = \overline{H(z,\xi)}\,.$$

and (3) is proved. Note that we also have

$$\sum_{\alpha=1}^{n} \frac{\partial G_\alpha}{\partial \bar{z}_\alpha}(\xi,z) = \overline{H(\xi,z)}\,. \tag{8.3}$$

To prove (4), recall that if $f \in C^1(\bar{D})$ is harmonic in D, we have

$$f(\xi) = \frac{-1}{2(n-1)w_{2n}} \int_{\partial D} f(z) \frac{\partial G(\xi,z)}{\partial n_z} dS_z \quad \text{for } \xi \in D\,. \tag{8.4}$$

Using complex notation as in Chapter 5, if $z \in \partial D$,

$$\frac{\partial G}{\partial n_z}\, dS_z = \frac{-i^n}{2^{n-2}} \sum_{\alpha=1}^{n} \frac{\partial G}{\partial z_\alpha}\, dz_\alpha \wedge dz_1 \wedge d\bar{z}_1 \wedge \ldots \widehat{dz_\alpha \wedge d\bar{z}_\alpha} \ldots \wedge dz_n \wedge d\bar{z}_n .$$

For $(\xi, z) \in D \times \partial D$ it follows from $G(\xi, z) = 0$ that $\frac{\partial G}{\partial \xi_\alpha}(\xi, z) = 0$; hence $G_\alpha(\xi, z) = \frac{\partial G}{\partial z_\alpha}(\xi, z)$ for $z \in \partial D$. Thus we may rewrite (8.4) as

$$f(\xi) = \frac{i^n}{2^{n-1}(n-1)w_{2n}} \sum_{\alpha=1}^{n} \int_{\partial D} f(z)\, G_\alpha(\xi, z) dz_\alpha \wedge dz_1 \wedge d\bar{z}_1 \wedge \ldots \widehat{dz_\alpha \wedge d\bar{z}_\alpha} \ldots \wedge dz_n \wedge d\bar{z}_n .$$

Using Stokes theorem together with (8.3), we obtain

$$\begin{aligned} f(\xi) &= \frac{i^n}{2^{n-1}(n-1)w_{2n}} \sum_{\alpha=1}^{n} \iint_D d[fG_\alpha dz_\alpha \wedge dz_1 \wedge d\bar{z}_1 \wedge \ldots \widehat{dz_\alpha \wedge d\bar{z}_\alpha} \ldots \wedge dz_n \wedge d\bar{z}_n] \\ &= \frac{-i^n}{2^{n-1}(n-1)w_{2n}} \sum_{\alpha=1}^{n} \iint_D \left(\frac{\partial f}{\partial \bar{z}_\alpha} G_\alpha + f \frac{\partial G_\alpha}{\partial \bar{z}_\alpha} \right) dz_1 \wedge d\bar{z}_1 \wedge \ldots \wedge dz_n \wedge d\bar{z}_n \\ &= \frac{-2}{(n-1)w_{2n}} \iint_D \left\{ \left[\sum_{\alpha=1}^{n} \frac{\partial f}{\partial \bar{z}_\alpha}(z)\, G_\alpha(\xi, z) \right] + f(z)\overline{H(\xi, z)} \right\} dV_z . \end{aligned}$$

for $\xi \in D$. ∎

An immediate consequence of Lemma 8.1 is the following formula (compare with the Bochner-Martinelli formula, cf. [K], Chapter 1).

<u>Corollary 8.1</u>. Let $f \in C^1(\bar{D})$ be holomorphic in D. Then for $\xi \in D$,

$$f(\xi) = \frac{-2}{(n-1)w_{2n}} (f, H(\xi, \cdot))_D .$$

To prove Theorem 8.1, we let $A(D)$ $(H(D))$ denote the space of holomorphic (harmonic) functions in $L^2(D)$. Clearly $A(D) \subset H(D)$, and, as

Hilbert spaces, we have the orthogonal decomposition

$$H(D) = A(D) \oplus A(D)^{\perp} .$$

Given $f \in H(D)$, we let f_A denote the projection of f to $A(D)$. By Corollary 8.1, we have

$$K(\xi,\cdot) = \frac{-2}{(n-1)w_{2n}} H(\xi,\cdot)_A$$

for $\xi \in D$. Hence

$$\text{(8.5)} \qquad K(\xi,\xi) \le \left(\frac{-2}{(n-1)w_{2n}}\right)^2 \|H(\xi,\cdot)\|_D^2 \quad \text{for } \xi \in D .$$

Proof of Theorem 8.1. Let $\beta \in \{1,2,\dots,n\}$ and $e_\beta = (0,\dots,\underset{\beta}{1},\dots,0)$.

By (8.1) applied to $a = e_\beta$, we obtain

$$\sum_{\beta=1}^{n} \frac{\partial^2(-\Lambda)}{\partial\xi_\beta\partial\bar{\xi}_\beta}(\xi) \ge \frac{4}{(n-1)w_{2n}} \sum_{\alpha,\beta=1}^{n} \left\| \frac{\partial G_\beta}{\partial\bar{z}_\alpha}(\xi,\cdot) \right\|_D^2$$

$$\ge \frac{4}{(n-1)w_{2n}} \sum_{\alpha=1}^{n} \left\| \frac{\partial G_\alpha}{\partial\bar{z}_\alpha}(\xi,\cdot) \right\|_D^2$$

$$\ge \frac{4}{(n-1)w_{2n}} \frac{1}{n} \left\| \sum_{\alpha=1}^{n} \frac{\partial G_\alpha}{\partial\bar{z}_\alpha}(\xi,\cdot) \right\|_D^2$$

$$= \frac{4}{n(n-1)w_{2n}} \left\| H(\xi,\cdot) \right\|_D^2$$

from (8.3). By (8.5), we thus have

$$\sum_{\beta=1}^{n} \frac{\partial^2(-\Lambda)}{\partial\xi_\beta\partial\bar{\xi}_\beta}(\xi) \ge \left(\frac{n-1}{n}\right) w_{2n} K(\xi,\xi)$$

for $\xi \in D$, which proves the theorem. ∎

Remark. If $T: D \to D_1$ is a biholomorphic mapping from D onto D_1, the Bergman kernel functions K, K_1 for D, D_1 transform according to the rule

$$K_1(w,w)|JT(\xi)|^2 = K(\xi,\xi)$$

where $w = T(\xi)$ and $JT(\xi)$ = Jacobian determinant of T. However, since harmonicity is not necessarily preserved under such mappings if $n \geq 2$, there is no general relationship between the Robin function $\Lambda(\xi)$ for (D,ξ) and $\Lambda_1(w)$ for (D_1, w). We also mention that Fefferman [F] proved that if $D \subset\subset \mathbb{C}^n$ is a strictly pseudoconvex domain with smooth (C^∞) boundary ∂D, then there exist $\phi, \tilde{\phi} \in C^\infty(\bar{D})$ with ϕ nonvanishing near ∂D such that

$$K(\xi,\xi) = \frac{\phi(\xi)}{(-\psi(\xi))^{n+1}} + \tilde{\phi}(\xi) \cdot \log(-\psi(\xi)), \quad \xi \in D, \tag{8.6}$$

where $D = \{\xi: \psi(\xi) < 0\}$, Grad $\psi(\xi) \neq 0$ for $\xi \in \partial D$, $\psi \in C^\infty(\bar{D})$ (i.e., ψ is a C^∞-defining function for D). From Lemma 3.1, it follows that $\Lambda(\xi)(-\psi(\xi))^{2n-2}$ is of class C^2 up to ∂D; hence, a natural question to ask is whether a formula analogous to (8.6) holds for $\Lambda(\xi)$ with $n + 1$ replaced by $2n - 2$.

9. METRIC INDUCED BY THE ROBIN FUNCTION

We return to a discussion of the 'metric' ds^2 introduced in Chapter 6 (cf. (6.11)). We let D be a bounded pseudoconvex domain in $\mathbb{C}^n$ with smooth boundary. From Corollary 7.1, $\log(-\Lambda(\xi))$ is a strictly plurisubharmonic function in D ; clearly $\log(-\Lambda(\xi))$ is also a real analytic exhaustion function for D, i.e., $\lim_{\xi\to\xi_0} \log(-\Lambda(\xi)) = +\infty$ for $\xi_0 \in \partial D$ (cf. [Y], Section 8). Thus the hermitian form

$$ds^2 \equiv \sum_{\alpha,\beta=1}^{n} \frac{\partial^2 \log(-\Lambda)}{\partial\xi_\alpha \partial\bar{\xi}_\beta}(\xi)\; d\xi_\alpha \otimes d\bar{\xi}_\beta$$

defines a Kähler metric in D which we call the <u>Λ-metric for D</u>. In the rest of this paper we study completeness of the Λ-metric for pseudoconvex domains D. The completeness criterion we use is the following (cf. [O]).

(cc) Let $ds_h^2 \equiv \sum_{\alpha,\beta=1}^{n} h_{\alpha\bar{\beta}}(\xi)\, d\xi_\alpha \otimes d\bar{\xi}_\beta$ be a Hermitian metric on D. Then ds_h^2 is <u>complete</u> in D if and only if for every C^1 curve γ: $t \to \xi(t)$, $t \in [0,1)$ in D that approaches ∂D as $t \longrightarrow 1^-$ (i.e., for any $K \subset D$ compact, $\exists\ 0 < t_0 < 1$ such that $\xi(t) \in D - K$ for all $t \in (t_0, 1)$) we have

$$\int_\gamma ds_h \equiv \lim_{t\to 1^-} \int_0^t \left[\sum_{\alpha,\beta=1}^{n} h_{\alpha\bar{\beta}}(\xi(t))\, \frac{d\xi_\alpha}{dt}(t)\, \overline{\frac{d\xi_\beta}{dt}}(t) \right]^{1/2} dt = +\infty .$$

A sufficient condition for (cc) is the following inequality relating the Levi-form of $\log(-\Lambda)$

(9.1) $$\ell_2(\xi,a) \equiv \sum_{\alpha,\beta=1}^{n} \frac{\partial^2 \log(-\Lambda)}{\partial\xi_\alpha \partial\bar{\xi}_\beta}(\xi)\, a_\alpha \bar{a}_\beta$$

and the complex directional derivative

(9.2) $$\ell_1(\xi,a) \equiv \sum_{\alpha=1}^{n} \frac{\partial \log(-\Lambda)}{\partial\xi_\alpha}(\xi)\, a_\alpha$$

where $\xi \in D$ and $a \in \mathbb{C}^n$. Note that $\ell_1(\xi,a) = \frac{c_1}{2}$ from Chapter 7.

<u>Lemma 9.1</u>. Let D be a pseudoconvex domain in $\mathbb{C}^n$ with smooth boundary. Suppose there exists a constant $c > 0$ such that

(9.3) $$\ell_2(\xi,a) \geq c|\ell_1(\xi,a)|^2$$

for all $\xi \in D$ and $a \in \mathbb{C}^n$. Then (cc) is satisfied for ds^2, i.e., the Λ-metric for D is complete.

<u>Proof</u>. Using (9.3) with $a = \xi'(t) = \left(\frac{d\xi_1}{dt}(t),\ldots,\frac{d\xi_n}{dt}(t)\right)$ where γ: $t \to \xi(t)$, $t \in [0,1)$, is a C^1 curve in D approaching ∂D, we have

$$\int_\gamma ds \geq \sqrt{c} \lim_{t\to 1^-} \int_0^t \left| \sum_{\alpha=1}^{n} \left[\frac{\partial \log(-\Lambda)}{\partial\xi_\alpha}(\xi(t)) \right] \frac{d\xi_\alpha(t)}{dt} \right| dt$$

$$\geq \frac{1}{2}\sqrt{c} \lim_{t\to 1^-} \left| \int_0^t \frac{d}{dt}\left[\log\left(-\Lambda(\xi(t))\right)\right] dt \right|$$

$$= \frac{1}{2}\sqrt{c} \lim_{t\to 1^-} \left| \log \frac{\Lambda(\xi(t))}{\Lambda(\xi(0))} \right| = +\infty . \ \blacksquare$$

Using a result in [Y], we show the following.

Theorem 9.1. Let D be a convex domain in $\mathbb{C}^n$ with smooth boundary. Then ds^2 is complete in D.

Proof. If we write $\xi_\alpha = x_{2\alpha} + ix_{2\alpha}$, $\alpha = 1,\ldots,n$, then formula 9.14 [Y] says that

$$(9.4) \qquad \sum_{i,j=1}^{2n} \frac{\partial^2 \log(-\Lambda)}{\partial x_i \partial x_j}(\xi)\, b_i b_j \geq \frac{1}{2n-2} \left| \sum_{j=1}^{2n} \frac{\partial \log(-\Lambda)}{\partial x_j}(\xi)\, b_j \right|^2$$

for $\xi \in D$ and $b = (b_1,\ldots,b_{2n}) \in R^{2n}$. We want to use Lemma 9.1. Since the inequality $\ell_2(\xi,a) \geq c|\ell_1(\xi,a)|^2$ is invariant under unitary tranformations of $\mathbb{C}^n$, it suffices to show that

$$(9.3)' \qquad \frac{\partial^2 \log(-\Lambda)}{\partial \xi_1 \partial \bar{\xi}_1}(\xi)\, |a|^2 \geq c \left| \frac{\partial^2 \log(-\Lambda)}{\partial \xi_1}(\xi)\, a \right|^2$$

for all $a \in \mathbb{C}$. But

$$\frac{\partial^2 \log(-\Lambda)}{\partial \xi_1 \partial \bar{\xi}_1}(\xi)\, |a|^2 = \frac{1}{4}\left[\frac{\partial^2 \log(-\Lambda)}{\partial x_1^2}(\xi)\, |a|^2 + \frac{\partial^2 \log(-\Lambda)}{\partial x_2^2}(\xi)\, |a|^2 \right]$$

$$\geq \frac{1}{4}\,\frac{1}{2n-2}\left[\left(\frac{\partial \log(-\Lambda)}{\partial x_1}(\xi)\, |a| \right)^2 + \left(\frac{\partial \log(-\Lambda)}{\partial x_2}(\xi)\, |a| \right)^2 \right]$$

(by (9.4))

$$= \frac{1}{2n-2} \left| \frac{\partial \log(-\Lambda)}{\partial \xi_1}(\xi)\, a \right|^2 . \blacksquare$$

For a general pseudoconvex domain $D = \{\xi \in \mathbb{C}^n\colon\ \psi(\xi) < 0\}$ with smooth boundary, in order to attempt to prove an inequality like (9.3), we analyze the formula (7.9) for $\tilde{J}(\xi,a)$, the volume integral in the computation of the Levi form $\ell_2(\xi,a)$ for $\log(-\Lambda(\xi))$. Recalling the definition of $H(a,\xi,z)$ in (7.10) and using the pseudoconvexity of D, we have the inequality

(9.5)

$$\ell_2(\xi,a) \geq \tilde{J}(\xi,a)$$

$$= \frac{4}{(n-1)w_{2n}(-\Lambda(\xi))}\iint_D \sum_{\alpha=1}^{n} \left|\frac{\partial}{\partial \bar{z}_\alpha}\left[-\ell_1(\xi,a)G_0(\xi,z) + \sum_{\beta=1}^{n} a_\beta G_\beta(\xi,z)\right]\right|^2 dV_z.$$

In this formula,

$$G_0(\xi,z) = G(\xi,z) + \frac{1}{n-1}\sum_{i=1}^{n}(z_i-\xi_i)\frac{\partial G}{\partial z_i}(\xi,z)$$

and

$$G_\alpha(\xi,z) = \frac{\partial G}{\partial \xi_\alpha}(\xi,z) + \frac{\partial G}{\partial z_\alpha}(\xi,z), \quad \alpha = 1,\ldots,n$$

for $(\xi,z) \in D \times D$. We want to express $\tilde{J}(\xi,a)$ in terms of $g(\xi,w)$ and its derivatives where $w = T_\xi(z) = \frac{z-\xi}{-\psi(\xi)}$ and $g(\xi,w)$ is the Green function of $(D(\xi),0)$. We had defined

$$g_0(\xi,w) = g(\xi,w) + \frac{1}{n-1}\sum_{i=1}^{n} w_i \frac{\partial g}{\partial w_i}(\xi,w) \quad \text{and}$$

(6.7)

$$g_\alpha(\xi,w) = \psi(\xi)\frac{\partial g}{\partial \xi_\alpha}(\xi,w) - (n-1)\frac{\partial \psi}{\partial \xi_\alpha}(\xi)\,(g_0(\xi,w) + \overline{g_0(\xi,w)})$$

in Chapter 6 and we had proved the relation

(6.9) $$g_\alpha(\xi,w) = \psi(\xi)^{2n-1}\,G_\alpha(\xi,z).$$

In a similar fashion, one can show that

(9.6) $$g_0(\xi,w) = \psi(\xi)^{2n-2}\,G_0(\xi,z).$$

Since $dV_w = \frac{1}{\psi(\xi)^{2n}} dV_z$, $\frac{\partial}{\partial \bar{z}_\alpha} = \frac{1}{-\psi(\xi)} \frac{\partial}{\partial \bar{w}_\alpha}$, and $\Lambda(\xi) = \frac{1}{\psi(\xi)^{2n-2}} \lambda(\xi)$

(where $\lambda(\xi)$ is the Robin constant of $(D(\xi),0)$) , we have

$$\tilde{J}(\xi,a) = \frac{4}{(n-1)w_{2n}(-\lambda(\xi))} \iint_{D(\xi)} \left(\sum_{\alpha=1}^{n} \left| \frac{\partial}{\partial \bar{w}_\alpha} \left(-\ell_1(\xi,a) g_0(\xi,w) + \frac{1}{\psi(\xi)} \sum_{\beta=1}^{n} a_\beta g_\beta(\xi,w) \right) \right|^2 \right) dV_w \tag{9.7}$$

We rewrite the integrand in a more convenient form. Differentiating $\lambda(\xi) = \psi(\xi)^{2n-2} \Lambda(\xi)$ with respect to ξ_α , we obtain

$$\frac{\partial \lambda}{\partial \xi_\alpha} = \frac{\partial \Lambda}{\partial \xi_\alpha} \psi^{2n-2} + (2n-2)\psi^{2n-3} \frac{\partial \psi}{\partial \xi_\alpha} \Lambda = \lambda \left[\frac{\frac{\partial \Lambda}{\partial \xi_\alpha}}{\Lambda} + (2n-2) \frac{\frac{\partial \psi}{\partial \xi_\alpha}}{\psi} \right] .$$

Hence

$$\frac{1}{\lambda} \sum_{\alpha=1}^{n} a_\alpha \frac{\partial \lambda}{\partial \xi_\alpha} = \ell_1 (\xi,a) + (2n-2) \sum_{\alpha=1}^{n} a_\alpha \frac{\frac{\partial \psi}{\partial \xi_\alpha}}{\psi} .$$

and we get that

(9.8)

$$h(a,\xi,w) \equiv -\ell_1(\xi,a) g_0(\xi,w) + \frac{1}{\psi(\xi)} \sum_{\alpha=1}^{n} a_\alpha g_\alpha (\xi,w)$$

$$= -\ell_1 \left(\frac{g_0 - \bar{g}_0}{2} \right) - \ell_1 \left(\frac{g_0 + \bar{g}_0}{2} \right) + \frac{1}{\psi} \sum_{\alpha=1}^{n} a_\alpha \left[\psi \frac{\partial g}{\partial \xi_\alpha} - (n-1) \frac{\partial \psi}{\partial \xi_\alpha} (g_0 + \bar{g}_0) \right]$$

$$= -\ell_1 \left(\frac{g_0 - \bar{g}_0}{2} \right) + \sum_{\alpha=1}^{n} a_\alpha \frac{\partial g}{\partial \xi_\alpha} - \left(\frac{g_0 + \bar{g}_0}{2} \right) \left(\sum_{\alpha=1}^{n} a_\alpha \frac{\frac{\partial \lambda}{\partial \xi_\alpha}}{\lambda} \right) .$$

Note that $h(a,\xi,w)$ is harmonic for $w \in D(\xi)$ by Proposition 6.2.

Theorem 9.2. Let $D \subset \mathbb{C}^n$ be a bounded pseudoconvex domain with smooth boundary. Then there exist constants $c, d > 0$ depending on D such that

$$[\ell_2(\xi,a)]^{1/2} \geq c|\ell_1(\xi,a)| - d \tag{9.9}$$

for all $\xi \in D$ and all $a \in \mathbb{C}^n$ with $\|a\| = 1$.

Proof. By (6.1), there exists $r_0 > 0$ such that $\{w: \|w\| < r_0\} \subset\subset D(\xi)$ for all $\xi \in \bar{D}$. Thus

$$\lambda(\xi) \geq -\frac{1}{r_0^{2n-2}}, \qquad \xi \in \bar{D}. \tag{9.10}$$

By (9.7), (9.8), (9.10), and the subharmonicity of $\left|\frac{\partial h}{\partial \bar{w}_\alpha}(a,\xi,w)\right|$ for $w \in D(\xi)$, we have

$$\ell_2(\xi,a) \geq \tilde{J}(\xi,a) \geq \frac{4}{(n-1)w_{2n}(-\lambda(\xi))} \iint_{\|w\|<r_0} \left(\sum_{\alpha=1}^{n} \left|\frac{\partial h}{\partial \bar{w}_\alpha}(a,\xi,w)\right|^2\right) dV_w$$

$$\geq \frac{4\Omega_{2n}\, r_0^{4n-2}}{(n-1)w_{2n}} \left(\sum_{\alpha=1}^{n} \left|\frac{\partial h}{\partial \bar{w}_\alpha}(a,\xi,0)\right|^2\right)$$

where Ω_{2n} = volume of the unit ball in $\mathbb{C}^n = \mathbb{R}^{2n}$. Hence

(9.11)

$$[\ell_2(\xi,a)]^{1/2} \geq \left(\frac{\Omega_{2n}}{(n-1)w_{2n}}\right)^{1/2} 2r_0^{2n-1} \left(\frac{1}{n}\left[\sum_{\alpha=1}^{n} \left|\frac{\partial h}{\partial \bar{w}_\alpha}(a,\xi,0)\right|\right]^2\right)^{1/2}$$

$$\geq \left(\frac{\Omega_{2n}}{n(n-1)w_{2n}}\right)^{1/2} 2r_0^{\,2n-1} \left\{ |\ell_1| \left[\sum_{\alpha=1}^{n} \left| \frac{\partial}{\partial \bar{w}_\alpha} \left(\frac{g_0 - \bar{g}_0}{2} \right) \right| \right] \right.$$

$$\left. - \sum_{\alpha=1}^{n} \left| \frac{\partial}{\partial \bar{w}_\alpha} \left[\sum_{\beta=1}^{n} a_\beta \frac{\partial g}{\partial \xi_\beta} - \left(\sum_{\beta=1}^{n} a_\beta \frac{\frac{\partial \lambda}{\partial \xi_\beta}}{\lambda} \right) \left(\frac{g_0 + \bar{g}_0}{2} \right) \right] \right| \right\}$$

where the term in { } is to be evaluated at $w = 0$. We estimate the terms on the right hand side of (9.11) as $\xi \to \partial D$. By Proposition 6.2 , the function

$$\frac{\partial}{\partial \bar{w}_\alpha} \left(\frac{g_0 - \bar{g}_0}{2} \right) \Bigg|_{(\xi,0)}$$

is continuous for $\xi \in \bar{D}$. From the explicit formula

(6.10)

$$g_0(\xi_0, w) = \frac{\sum_{\beta=1}^{n} (\bar{N}_{\xi_0})_\beta \bar{w}_\beta - \|N_{\xi_0}\|^2}{\|w - (\bar{N}_{\xi_0})\|^{2n}} \quad \text{for } \xi_0 \in \partial D \text{ where } N_{\xi_0} = \frac{\mathrm{Grad}\psi(\xi_0)}{\|\mathrm{Grad}\psi(\xi_0)\|^2} ,$$

it follows that

$$\frac{g_0(\xi_0, w) - \bar{g}_0(\xi_0, w)}{2} = \frac{1}{2} \frac{\sum_{\beta=1}^{n} \left(\bar{w}_\beta \frac{\partial \psi}{\partial \bar{\xi}_\beta}(\xi_0) - w_\beta \frac{\partial \psi}{\partial \xi_\beta}(\xi_0) \right)}{\|w - (\bar{N}_{\xi_0})\|^{2n} \|\mathrm{Grad}\psi(\xi_0)\|^2}$$

Hence for $\xi \in D$, $\xi_0 \in \partial D$,

$$\lim_{\xi \to \xi_0} \sum_{\alpha=1}^{n} \left| \frac{\partial}{\partial \bar{w}_\alpha} \left(\frac{g_0 - \bar{g}_0}{2} \right) \right|_{(\xi,0)} = \sum_{\alpha=1}^{n} \left| \frac{\partial}{\partial \bar{w}_\alpha} \left(\frac{g_0 - \bar{g}_0}{2} \right) \right|_{(\xi_0,0)}$$

$$= \frac{1}{2} \|\mathrm{Grad}\ \psi(\xi_0)\|^{2n-2} \left(\sum_{\alpha=1}^{n} \left| \frac{\partial \psi}{\partial \bar{\xi}_\alpha}(\xi_0) \right| \right) > 0$$

since ψ is a smooth defining function for D. If we set

$$A \equiv \frac{1}{3} \min_{\xi\in\partial D} \left\{ \|\text{Grad } \psi(\xi)\|^{2n-2} \left(\sum_{\alpha=1}^{n} \left| \frac{\partial\psi}{\partial\bar{\xi}_\alpha} (\xi) \right| \right) \right\} > 0 ,$$

then by compactness of ∂D, there exists a neighborhood U of ∂D such that

$$\sum_{\alpha=1}^{n} \left| \frac{\partial}{\partial\bar{w}_\alpha} \left(\frac{g_0 - \bar{g}_0}{2} \right) \right|_{(\xi,0)} > A \tag{9.12}$$

for all $\xi \in D \cap U$.

To estimate the second term on the right hand side of (9.11), note that from Corollary 4.1 and Proposition 6.2 it follows that the functions

$\dfrac{\partial^2 g}{\partial\bar{w}_\alpha \partial\xi_\beta} (\xi,0)$, $\dfrac{\partial}{\partial\bar{w}_\alpha}\left(\dfrac{g_0+\bar{g}_0}{2} \right)(\xi,0)$ and $\dfrac{\partial\lambda}{\partial\xi_\beta} (\xi)$ are certainly continuous

for $\xi \in \bar{D}$. This fact, together with (9.10), imply the existence of a constant $B > 0$ such that

$$\sum_{\alpha=1}^{n} \left| \frac{\partial}{\partial\bar{w}_\alpha} \left[\sum_{\beta=1}^{n} a_\beta \frac{\partial g}{\partial\xi_\beta} - \left(\sum_{\beta=1}^{n} a_\beta \frac{\frac{\partial\lambda}{\partial\xi_\beta}}{\lambda} \right) \left(\frac{g_0+\bar{g}_0}{2} \right) \right] \right|_{(\xi,0)} < B \tag{9.13}$$

for all $\xi \in \bar{D}$ and $a \in \mathbb{C}^n$ with $\|a\| = 1$. Note that we are again using compactness of $\bar{D}$.

From (9.12), (9.13) and (9.11), we have

$$\left[\ell_2(\xi,a)\right]^{1/2} \geq \left(\frac{\Omega_{2n}}{n(n-1)w_{2n}} \right)^{1/2} 2r_0^{\,2n-1} \left[A|\ell_1(\xi,a)| - B \right]$$

for $\xi \in D \cap U$ and $a \in \mathbb{C}^n$ with $\|a\| = 1$. From Corollary 7.1,

$\ell_2(\xi, a) > 0$ on D if $a \neq 0$; thus, since $D - U$ is compact and $\ell_1(\xi, a)$, $\ell_2(\xi, a)$ are continuous for $\xi \in D - U$ and $\|a\| = 1$, there exists a constant A' such that

$$\left[\ell_2(\xi, a)\right]^{1/2} > A' |\ell_1(\xi, a)|$$

for $\xi \in D - U$, $\|a\| = 1$. Setting $c = \min\left\{\left(\frac{\Omega_{2n}}{n(n-1)w_{2n}}\right)^{1/2} 2r_0^{2n-1} A, A'\right\}$

and $d = \left(\frac{\Omega_{2n}}{n(n-1)w_{2n}}\right)^{1/2} 2r_0^{2n-1} B$ gives (9.9) and proves the theorem. ∎

Theorem 9.2 can be used to show that a large class of curves tending to ∂D have infinite length in the Λ-metric.

Corollary 9.1. Let D be a pseudoconvex domain in $\mathbb{C}^n$ with smooth boundary. Then

$$\int_\gamma ds = + \infty$$

for any C^1 curve γ: $t \to \xi(t)$, $0 \le t < 1$, in D approaching ∂D which has finite Euclidean length.

Proof. We parameterize γ: $\sigma \to \xi(\sigma)$, $0 \le \sigma < L$, using Euclidean arclength; our assumption is that $L < + \infty$. From the theorem,

$$\int_\gamma ds \ge \lim_{\sigma \to L^-} \int_0^\sigma \left\{ c \left| \sum_{\alpha=1}^n \frac{\partial \log(-\Lambda)}{\partial \xi_\alpha} (\xi(\sigma)) \frac{d\xi_\alpha}{d\sigma} (\sigma) \right| - d \right\} d\sigma$$

$$\ge \frac{c}{2} \lim_{\sigma \to L^-} \left| \log \frac{\Lambda(\xi(\sigma))}{\Lambda(\xi(0))} \right| - dL = + \infty . \ ∎$$

Remark. Based on Theorem 6.1 and Corollary 9.1, it seems likely that the Λ-metric for a bounded pseudoconvex domain $D \subset \mathbb{C}^n$ with smooth boundary is complete, but we cannot prove this.

We continue to analyze (9.3). First, we introduce the notation

$$S_n \equiv \{a \in \mathbb{C}^n : \|a\| = 1\}$$

and, for $\rho > 0$, we set

$$M_\rho \equiv \{(\xi, a) \in D \times S_n : |\ell_1(\xi, a)| < \rho\} ;$$

$$M'_\rho \equiv \{(\xi, a) \in D \times S_n : |\ell_1(\xi, a)| \geq \rho\} .$$

For each fixed $a \in S_n$,

$$M_\rho(a) \equiv \{\xi \in D : (\xi, a) \in M_\rho\}$$

is a set of points in D where the complex gradient of $\log(-\Lambda)$ is 'almost' orthogonal to $\bar{a}$. Using Theorem 9.2, we can get, for certain $\rho > 0$, an estimate as in (9.3) for pairs $(\xi, a) \in M'_\rho$.

Corollary 9.2. Under the same hypothesis as in Theorem 9.2, let $\rho_0 \equiv \frac{2d}{c} > 0$. Then

$$\ell_2(\xi, a) \geq \left(\frac{c}{2}\right)^2 |\ell_1(\xi, a)|^2 \tag{9.14}$$

for all $(\xi, a) \in M'_{\rho_0}$.

Proof. If $(\xi, a) \in M'_{\rho_0}$, then $|\ell_1(\xi, a)| \geq \rho_0$. Thus by Theorem 9.2,

$$[\ell_2(\xi, a)]^{1/2} > c\,|\ell_1(\xi, a)| - d = c|\ell_1(\xi, a)| - \frac{c\rho_0}{2} \geq \frac{c}{2}|\ell_1(\xi, a)|$$

which is (9.14). ■

Remark. In Chapter 13 we construct a bounded pseudoconvex domain $D \subset \mathbb{C}^n$ with smooth boundary for which there exists no positive constant c such that (9.3) holds for all $(\xi, a) \in D \times S_n$.

10. STRICTLY PSEUDOCONVEX BOUNDARY POINTS

In this chapter, we exploit the existence of strictly pseudoconvex boundary points on the boundary of a bounded, smoothly bounded domain $D \subset \mathbb{C}^n$, and, using the explicit formula (7.14) for $\ell_2(\xi, a)$, we examine properties of the Λ-metric. Our first result is a quantitative statement about the behavior of $\ell_2(\xi, a)$ for $(\xi, a) \in M_\rho$ if ξ is near a strictly pseudoconvex boundary point.

<u>Lemma 10.1</u>. Let D be a bounded pseudoconvex domain in $\mathbb{C}^n$ with smooth boundary and let ψ be a smooth defining function for D. Let $z_0 \in \partial D$ be a strictly pseudoconvex boundary point and let $\rho > 0$. Then there exist a neighborhood V of z_0 and a constant $c > 0$ such that

$$\ell_2(\xi, a) \geq \frac{c}{-\psi(\xi)}$$

for all $(\xi, a) \in M_\rho$ with $\xi \in D \cap V$.

<u>Proof</u>. Fix $z_0 \in \partial D$ and $\rho > 0$ as in the hypothesis. By Proposition 7.1, since z_0 is a strictly pseudoconvex boundary point, there exists $m > 0$ such that

$$K_2(z_0, a) > m \quad \text{for all} \quad a \in S_n .$$

Setting $c_1 = 2\ell_1(\xi, a) = 2 \sum_{\alpha=1}^{n} \frac{\partial \log(-\Lambda)}{\partial \xi_\alpha} (\xi) a_\alpha$ as in Chapter 7, we can clearly find a neighborhood V_1 of z_0 such that

$$\|\mathcal{O}(a, \xi, z_0)\| = \left\| \frac{-c_1}{2n-2} (z_0 - \xi) + a \right\| \geq \frac{2}{3}$$

for all $(\xi, a) \in M_\rho$ with $\xi \in V_1$. Then we can find a perhaps smaller neighborhood $V_2 \subset V_1$ such that, if $\gamma = V_2 \cap \partial D$,

(10.1) $$\|\mathcal{O}(a,\xi,z)\| \geq \frac{1}{2} \quad \text{for} \quad (\xi,a) \in M_\rho\ ,\ \xi \in V_2\ ,\ z \in \gamma$$

and, by continuity,

(10.2) $$K_2(z,a) \geq \frac{m}{2} \quad \text{for} \quad (z,a) \in \gamma \times S_n$$

We use (10.1) and (10.2) to estimate $K_2(z,\mathcal{O})$ if $z \in \gamma$. By property (1) of K_2 in Chapter 7,

$$K_2(z,\mathcal{O}) = \|\mathcal{O}\|^2\ K_2(z,\frac{\mathcal{O}}{\|\mathcal{O}\|}) .$$

Thus for $(\xi,a) \in M_\rho$, $\xi \in V_2$ and $z \in \gamma$,

(10.3) $$K_2(z,\mathcal{O}) \geq \frac{1}{4}\,\frac{m}{2} = \frac{m}{8} .$$

Since D is pseudoconvex, $K_2(z,\mathcal{O}) \geq 0$ for all $(z,\mathcal{O}) \in \partial D \times \mathbb{C}^n$, hence, from (7.13) and (7.14) and our above estimates,

(10.4) $$\ell_2(\xi,a) \geq \tilde{I}(\xi,a) \geq \frac{1}{(n-1)w_{2n}(-\Lambda(\xi))} \int_\gamma K_2(z,\mathcal{O})\|\text{Grad}_{(z)}G(\xi,z)\|^2 dS_z$$

$$\geq \frac{1}{(n-1)w_{2n}(-\Lambda(\xi))}\,\frac{m}{8} \int_\gamma \|\text{Grad}_{(z)}G(\xi,z)\|^2 dS_z$$

if $(\xi,a) \in M_\rho$.

To estimate the integral in (10.4), we again consider the transformation $w = T_\xi(z) = \frac{z-\xi}{-\psi(\xi)}$ for $\xi \in D$. We set

$$\Gamma(\xi) \equiv \begin{cases} T_\xi(\gamma) & \text{if } \xi \in D \\ \partial D(\xi) & \text{if } \xi \in \partial D \end{cases}$$

where we recall that for $\xi \in \partial D$,

$$\partial D(\xi) = \left\{ w \in \mathbb{C}^n:\ 2\,\text{Re}\left(\sum_{\alpha=1}^{n} \frac{\partial\psi}{\partial\xi_\alpha}(\xi) w_\alpha \right) - 1 = 0 \right\}$$

is a real hyperplane in $\mathbb{C}^n$. As $\xi \in D$ approaches $z_0 \in \partial D$, the boundary surfaces $\Gamma(\xi)$ approach $\Gamma(z_0)$. More precisely, given $R > 1$, we can find a neighborhood U of z_0 such that the variation

$$\bar{\mathcal{D}}\Big|_{R,U} : \quad \xi \to \overline{D(\xi)} \cap \{w: \|w\| < R\}, \quad \xi \in \bar{D} \cap U,$$

is diffeomorphically equivalent to a trivial variation (cf. Proposition 4.2 [Y]) so that if $\xi \in D \cap U$ approaches $z_0 \in \partial D$, the surfaces

$$\Gamma(\xi) \cap \{w: \|w\| < R\}$$

approach

$$\partial D(z_0) \cap \{w: \|w\| < R\}$$

in such a way that the outer normal vectors $\mathrm{Grad}_{(w)} g(\xi, w)$ approach $\mathrm{Grad}_{(w)} g(z_0, w)$ in a continuous fashion. Rewriting (10.4) using the relationship $g(\xi, w) = \psi(\xi)^{2n-2}\, G(\xi, z)$ and $\lambda(\xi) = \psi(\xi)^{2n-2} \Lambda(\xi)$, we obtain

$$\ell_2(\xi, a) \geq \frac{m}{8(n-1)w_{2n}} \frac{(\psi(\xi))^{2n-2}}{(-\lambda(\xi))} \int_{\Gamma(\xi)} \left[\frac{1}{\psi(\xi)^{2n-1}} \|\mathrm{Grad}_{(w)} g(\xi, w)\|\right]^2$$

$$\cdot\ (-\psi(\xi))^{2n-1}\, dS_w \tag{10.5}$$

$$= \frac{m}{8(n-1)w_{2n}} \frac{1}{(-\lambda(\xi))} \frac{1}{(-\psi(\xi))} \int_{\Gamma(\xi)} \|\mathrm{Grad}_{(w)} g(\xi, w)\|^2\, dS_w$$

From the above remarks about $\mathrm{Grad}_{(w)} g(\xi, w)$ and the continuity of $\lambda(\xi)$ up to ∂D, by Fatou's lemma, we have

$$\varliminf_{\substack{\xi \to z_0 \\ \xi \in D}} \frac{1}{(n-1)w_{2n}} \frac{1}{(-\lambda(\xi))} \int_{\Gamma(\xi)} \|\mathrm{Grad}_{(w)} g(\xi, w)\|^2\, dS_w$$

$$\geq \frac{1}{(n-1)w_{2n}} \frac{1}{\|\mathrm{Grad}\psi(z_0)\|^{2n-2}} \int_{\Gamma(z_0)} \|\mathrm{Grad}_{(w)} g(z_0, w)\|^2\, dS_w$$

$$= 2(n-1)\, \|\mathrm{Grad}\ \psi(z_0)\| > 0 .$$

This last equality is an explicit calculation which will be performed in Chapter 11 (cf. (11.12)). Thus we can find a neighborhood $V \subset V_2$ of z_0 such that

$$\frac{1}{(n-1)w_{2n}} \frac{1}{(-\lambda(\xi))} \int_{\Gamma(\xi)} \|\mathrm{Grad}_{(w)} g(\xi, w)\|^2 \, dS_w > (n-1) \|\mathrm{Grad}\, \psi(z_0)\|$$

for all $\xi \in D \cap V$. Thus (10.5) and the above inequality yield that

$$\ell_2(\xi, a) \ge \frac{m}{8} (n-1) \|\mathrm{Grad}\, \psi(z_0)\| \left[\frac{-1}{\psi(\xi)} \right] \equiv \frac{C}{-\psi(\xi)}$$

for all $(\xi, a) \in M_\rho$ with $\xi \in D \cap V$. ∎

Combining Lemma 10.1 with Corollary 9.1, we have the following.

Corollary 10.1. Let D be a bounded pseudoconvex domain in $\mathbb{C}^n$ with smooth boundary. Let $z_0 \in \partial D$ be a strictly pseudoconvex boundary point. Then there exists a neighborhood U of z_0 and a constant $\tilde{c} > 0$ such that

$$\ell_2(\xi, a) \ge \tilde{c} \, |\ell_1(\xi, a)|^2 \tag{10.6}$$

for all $\xi \in D \cap U$, $a \in S_n$.

Proof. By Corollary 9.1, we have constants $c, d > 0$ and $\rho_0 \equiv 2d/c$ such that

$$\ell_2(\xi, a) \ge \left(\frac{c}{2} \right)^2 |\ell_1(\xi, a)|^2$$

for all $(\xi, a) \in M'_{\rho_0}$. Setting $\rho = \rho_0$ in Lemma 10.1, we find a neighborhood V of z_0 and $C > 0$ such that

$$\ell_2(\xi, a) \ge \frac{C}{-\psi(\xi)}$$

for all $(\xi, a) \in M_{\rho_0}$ with $\xi \in D \cap V$. Since $\lim_{\xi \to z_0} \psi(\xi) = \psi(z_0) = 0$, we can find a neighborhood $V_0 \subset V$ of z_0 with

$$\frac{1}{-\psi(\xi)} > \rho_0^2$$

for $\xi \in V_0 \cap D$. Thus, for $(\xi, a) \in M_{\rho_0}$ with $\xi \in V_0$,

$$\ell_2(\xi, a) \geq \frac{C}{-\psi(\xi)} > C\rho_0^2 > C|\ell_1(\xi, a)|^2$$

by definiton of M_{ρ_0}. Setting $U = V_0$ and $\tilde{c} = \min\left\{ (\frac{c}{2})^2, C\right\}$ proves the Corollary. ∎

<u>Corollary</u> <u>10.2</u>. Let D be a bounded pseudoconvex domain in $\mathbb{C}^n$ with smooth boundary. Let z_0, U be as in Corollary 10.1. Then for any C^1 curve γ: $t \to \xi(t)$ $(0 \leq t < 1)$ which tends to a boundary point $\xi_0 \in U \cap \partial D$ (i.e., $\lim_{t\to 1^-} \xi(t) = \xi_0 \in \partial D$), $\int_\gamma ds = +\infty$.

<u>Proof</u>. Since γ tends to $\xi_0 \in U \cap \partial D$, there exists $0 < t_0 < 1$ with $\xi(t) \in U \cap \bar{D}$ for $t > t_0$. Hence

$$\int_\gamma ds \geq \int_{\gamma\cap U} ds \geq \lim_{t\to 1^-} \int_{t_0}^{t} [\ell_2(\xi(t), \xi'(t))]^{1/2}\, dt$$

$$\geq \lim_{t\to 1^-} \frac{(\tilde{c})^{1/2}}{2} \left| \int_{t_0}^{t} \frac{d}{dt} \log\left(-\Lambda(\xi(t))\right) dt \right| \quad \text{(by (10.6))}$$

$$= \frac{(\tilde{c})^{1/2}}{2} \lim_{t\to 1^-} \left| \log \frac{\Lambda(\xi(t))}{\Lambda(\xi(t_0))} \right| = +\infty . \quad ∎$$

<u>Corollary</u> <u>10.3</u>. Let D be a bounded strictly pseudoconvex domain in $\mathbb{C}^n$ with smooth boundary. Then

(a) for any $a \in S_n$ and $\xi_0 \in \partial D$, $\lim_{\substack{\xi\to\xi_0 \\ \xi\in D}} \ell_2(\xi, a) = +\infty$

(b) there exists $c > 0$ such that

$$\ell_2(\xi, a) \geq c|\ell_1(\xi, a)|^2$$

for all $(\xi, a) \in D \times S_n$.

(c) the Λ-metric is complete in D .

Proof. (a) If the conclusion is false, then we can choose a sequence $\{\xi^j\} \subset D$ and $a \in S_n$ with

$$\lim_{j\to+\infty} \xi^j = \xi_0 \quad \text{and} \quad \lim_{j\to+\infty} \ell_2(\xi^j, a) < +\infty .$$

We distinguish two cases:

If $\overline{\lim}_{j\to+\infty} |\ell_1(\xi^j, a)| = +\infty$, then by Theorem 9.2, there exist constants $c, d > 0$ so that

$$[\ell_2(\xi^j, a)]^{1/2} \geq c|\ell_1(\xi^j, a)| - d$$

for all $j = 1, 2, \ldots$ which implies that

$$\overline{\lim}_{j\to+\infty} \left[\ell_2(\xi^j, a)\right]^{1/2} = +\infty ,$$

contradicting our assumption.

If $\overline{\lim}_{j\to+\infty} \left|\ell_1(\xi^j, a)\right| < +\infty$, let $\rho \equiv \sup_j \left|\ell_1(\xi^j, a)\right| < +\infty$. By Lemma 10.1 , since each $(\xi^j, a) \in M_\rho$, we get a $c > 0$ such that

$$\lim_{j\to+\infty} \ell_2(\xi^j, a) \geq c \lim_{j\to+\infty} \left[\frac{-1}{\psi(\xi^j)}\right] = +\infty$$

since $\lim_{j\to+\infty} \xi^j = \xi_0 \in \partial D$. This again contradicts our assumption and (a) is proved.

(b), (c) We show that there exists $c > 0$ such that

(9.3) $$\ell_2(\xi,a) \geq c|\ell_1(\xi,a)|^2$$

for all $(\xi,a) \in D \times S_n$; by Lemma 9.1 it follows that the Λ-metric is complete. By Corollary 10.1, for each $\xi_0 \in \partial D$ we can find a neighborhood U_{ξ_0} of ξ_0 and a constant $c_{\xi_0} > 0$ such that

$$\ell_2(\xi,a) \geq c_{\xi_0}|\ell_1(\xi,a)|^2$$

for all $(\xi,a) \in (D \cap U_{\xi_0}) \times S_n$. By compactness of ∂D , a finite collection $U_{\xi_1},\ldots,U_{\xi_m}$ of these sets cover ∂D . Thus $U \equiv U_{\xi_1} \cup\ldots\cup U_{\xi_m}$ is a neighborhood of ∂D and

$$\ell_2(\xi,a) \geq c'|\ell_1(\xi,a)|^2$$

for $(\xi,a) \in U \times S_n$ where $c' = \min(c_{\xi_1},\ldots,c_{\xi_m})$. Since $D - U$ is compact and $\ell_2(\xi,a)$, $\ell_1(\xi,a)$ are continuous on $(D-U) \times S_n$, we can find $c'' > 0$ such that

$$\ell_2(\xi,a) \geq c''|\ell_1(\xi,a)|^2$$

for $(\xi,a) \in (D-U) \times S_n$. Setting $c = \min(c',c'')$ yields (9.3) for all $(\xi,a) \in D \times S_n$. ∎

<u>Remark.</u> In (a), the result clearly holds if we allow a sequence $\{a_j\}$ with $\lim_{j\to+\infty} a_j = a$ in the hypothesis.

To illustrate the difference in the behavior of the Λ-metric for strictly pseudoconvex domains versus weakly pseudoconvex domains, we introduce a notion of point separation by a metric. Let D be a domain in $\mathbb{C}^n$ and let ds^2 be an arbitrary metric on D . We say that (D,ds^2) has the <u>separation property</u> (resp., <u>strict separation property</u>) if, given any $\xi',\xi'' \in \partial D$ with $\xi' \neq \xi''$, and given any sequences $\{\xi'_j\}$, $\{\xi''_j\} \subset D$ with

$$\lim_{j\to+\infty} \xi'_j = \xi' \quad \text{and} \quad \lim_{j\to+\infty} \xi''_j = \xi'' ,$$

we have

(10.7)

$$\varliminf_{j\to+\infty} \left[\inf\left\{ \int_{C(\xi'_j,\xi''_j)} ds : C(\xi'_j,\xi''_j) \text{ is a } C^1 \text{ curve joining } \xi'_j,\xi''_j \right\} \right] > 0$$

(resp., $+\infty$) .

Part (a) of Corollary 10.3 immediately yields the following result.

Corollary 10.4. Let D be a bounded strictly pseudoconvex domain in $\mathbb{C}^n$ with smooth boundary. Then (D, ds^2) has the strict separation property where ds^2 is the Λ-metric for D .

Corollary 10.3 (a) and Corollary 10.4 are not true in general if D is only weakly pseudoconvex as the following example illustrates. Let

$$B = \{z \in \mathbb{C}^n : \|z\| < 1\} , \quad \Omega = \{z \in \mathbb{C}^n : \ \operatorname{Im} z_n > 0\} ,$$

and set $D = B \cap \Omega$. By modifying D near $\partial B \cap \partial\Omega$, we can construct a smoothly bounded convex domain D' which is Levi-flat at $z = 0$. By Theorem 9.1, inequality (9.3) holds for D' and the Λ-metric ds^2 is complete for D' . However, for $a = (1,0,\ldots,0)$ and $\xi = 0$, we show that

(10.8) $$\lim_{\substack{\xi\to 0 \\ \xi\in D}} \ell_2(\xi, a) = 0$$

and thus (10.8) remains valid for D' . Thus (a) of Corollary (10.3) is not satisfied; furthermore, by taking ξ' and ξ'' sufficiently close to $\xi = 0$ and using (10.8) for D' , we see that (D', ds^2) does not have the separation property.

To prove (10.8), we let $G(\xi,z)$ and $\Lambda(\xi)$ be the Green function and Robin constant for (D,ξ) and we let $G_B(\xi,z)$, $\Lambda_B(\xi)$ be the corresponding ones for (B,ξ). By symmetry, for $\xi = (\xi_1,\ldots,\xi_n) \in D$,

$$G(\xi,z) = G_B(\xi,z) - G_B(\tilde{\xi},z)$$

where $\tilde{\xi} = (\xi_1,\ldots,\xi_{n-1},\bar{\xi}_n)$. Recalling that

$$G_B(\xi,z) = \frac{1}{\|z-\xi\|^{2n-2}} - \frac{1}{\|\xi\|^{2n-2}} \frac{1}{\left\|z - \frac{\xi}{\|\xi\|^2}\right\|^{2n-2}},$$

it follows that

$$\begin{aligned}\Lambda(\xi) &= \Lambda_B(\xi) - G_B(\tilde{\xi},\xi) \\ &= \frac{-1}{(1-\|\xi\|^2)^{2n-2}} - \left\{ \frac{1}{\|\xi-\tilde{\xi}\|^{2n-2}} - \frac{1}{\|\tilde{\xi}\|^{2n-2}} \frac{1}{\left\|\xi - \frac{\tilde{\xi}}{\|\tilde{\xi}\|^2}\right\|^{2n-2}} \right\}.\end{aligned}$$

Since

$$\frac{1}{\|\tilde{\xi}\|^{2n-2}} \frac{1}{\left\|\xi - \frac{\tilde{\xi}}{\|\tilde{\xi}\|^2}\right\|^{2n-2}} = \frac{1}{(1-\|\tilde{\xi}\|^2)^{2n-2}} + H(\tilde{\xi},\xi)$$

where $H(\tilde{\xi},\xi)$ is regular for $(\tilde{\xi},\xi) \in B \times B$ and satisfies $H(\tilde{\xi},\xi) = 0$, it follows that

$$\Lambda(\xi) = \frac{-1}{|\xi_n-\bar{\xi}_n|^{2n-2}} + H(\tilde{\xi},\xi).$$

Thus, since

$$\ell_2(\xi, a) = \frac{\partial^2 \log(-\Lambda)}{\partial\xi_1 \partial\bar{\xi}_1}(\xi) \quad \text{for} \quad a = (1,0,\dots,0),$$

$$\lim_{\substack{\xi\to 0\\ \xi\in D}} \ell_2(\xi, a) = \lim_{\substack{\xi\to 0\\ \xi\in D}} \frac{\partial^2 \log[1-|\xi_n-\bar{\xi}_n|^{2n-2}H(\tilde{\xi},\xi)]}{\partial\xi_1\partial\bar{\xi}_1} = 0$$

and (10.8) is proved. Thus ds^2 does not separate points ξ', ξ'' near 0 on $\partial D' \cap \partial\Omega$.

Remark. The smoothness assumption on ∂D is clearly necessary in order to have a complete Λ-metric. To show this, we first of all consider the Hartogs triangle

$$D \equiv \{(z_1, z_2) \in \mathbb{C}^2 \colon \ |z_2| < |z_1|\}$$

and set

$$D_R = D \cap B_R \equiv D \cap \{(z_1, z_2) \in \mathbb{C}^2\colon \ |z_1|^2 + |z_2|^2 < R^2\}$$

for $R > 1$. Then D_R is a domain with corners

$$e \equiv \partial D \cap \partial B_R \ ;$$

clearly we can modify D_R near e to obtain a pseudoconvex domain D'_R whose boundary $\partial D'_R$ is smooth except at the point $(0,0)$. Furthermore, we can assume that, e.g.,

$$D_{R/2} \subset D'_R \ .$$

Let $\Lambda(\xi)$ be the Robin function for D and $\Lambda_R(\xi)$ the Robin function for D'_R . The existence of strictly pseudoconvex boundary points on $\partial D'_R$ implies that $\log(-\Lambda_R(\xi))$ is strictly plurisubharmonic in D'_R ; hence the Λ_R-metric is a Kähler metric. We show, however, that this metric is not complete in D'_R .

Since D is invariant under a scaling

$$T(z_1, z_2) = (cz_1, cz_2), \quad c \neq 0 ,$$

it follows that

$$\Lambda(t,0) = \frac{1}{|t|^2} \Lambda(1,0) \quad \text{for} \quad t \neq 0 .$$

Since $D_{R/2} \subset D_{R'}$, one can show, using elementary arguments, that

$$\Lambda_R(t,0) = \Lambda(t,0) - o(|t|) \quad \text{as} \quad |t| \to 0 .$$

Hence

$$\log(-\Lambda_R(t,0)) = -2 \log|t| + \log[-\Lambda(1,0) - o(|t|)|t|^2] .$$

Since $\frac{\partial^2}{\partial t \partial \bar{t}} \log |t|^2 = 0$ for $t \neq 0$ and $-\Lambda(1,0) > 0$, it follows that

$$\frac{\partial^2 \log(-\Lambda_R)}{\partial z_1 \partial \bar{z}_1} (t,0) = o(|t|) \quad \text{as} \quad |t| \to 0 .$$

Thus, taking the curve γ: $\tau \to \xi(\tau) \equiv (1-\tau, 0)$, $0 \leq \tau < 1$, we have, in the Λ_R-metric ds_R^2 ,

$$\int_\gamma ds_R = \lim_{\tau \to 1^-} \int_0^\tau \left[\frac{\partial^2 \log(-\Lambda_R)}{\partial z_1 \partial \bar{z}_1} (1-\tau, 0) \right]^{1/2} d\tau$$

$$= \lim_{\tau \to 1^-} \int_0^\tau [o(|1-\tau|)]^{1/2} d\tau < +\infty .$$

By (CC), the Λ_R-metric is not complete in D'_R . We remark that in this example,

$$\lim_{\substack{\xi \to \xi_0 \\ \xi \in D'_R}} \log(-\Lambda_R(\xi)) = +\infty$$

for each $\xi_0 \in \partial D'_R$, even $\xi_0 = (0,0)$, since each such point is regular for the Dirichlet problem (c.f., [Y], Section 8, (2)).

11. EXPLICIT FORMULAS FOR A HALF-SPACE

In this chapter, we let $D = \{z \in \mathbb{C}^n:\ \psi(z) < 0\}$ be a bounded pseudoconvex domain in $\mathbb{C}^n$ with smooth boundary. Following our usual notation, we let

$$D(\xi) = \left\{ w = \frac{z-\xi}{-\psi(\xi)} \ :\ z \in D \right\}$$

for each $\xi \in D$. Recall that as $\xi \in D$ approaches a boundary point $\xi_0 \in \partial D$, the domains $D(\xi)$ approach the half-space

$$D(\xi_0) \equiv \left\{ w \in \mathbb{C}^n:\ 2\,\mathrm{Re}\left\{ \sum_{\alpha=1}^{n} w_\alpha \frac{\partial\psi}{\partial z_\alpha}(\xi_0) \right\} - 1 \equiv f(\xi_0, w) < 0 \right\} .$$

We want to compute some of the limiting quantities involving the Robin function $\Lambda(\xi)$ when ξ tends to ξ_0. In particular, we analyze the behavior of the ratio

$$\ell_2(\xi, a)/|\ell_1(\xi, a)|^2 \ , \quad a \in S_n \ ,$$

as $\xi \to \xi_0$. This will require some explicit computations involving the half-space, $D(\xi_0)$.

Recall that $0 \in D(\xi_0)$ so that we have the explicit formula for the Green function $g(\xi_0, w)$ for $(D(\xi_0), 0)$,

$$g(\xi_0, w) = \frac{1}{\|w\|^{2n-2}} - \frac{1}{\|w - \bar{N}_{\xi_0}\|^{2n-2}} \tag{11.1}$$

where $\bar{N}_{\xi_0} = \bar{N} = \dfrac{\overline{\mathrm{Grad}\psi(\xi_0)}}{\|\mathrm{Grad}\psi(\xi_0)\|^2}$ is the symmetric point of the origin with respect to the hyperplane

$$\partial D(\xi_0) = \left\{ w \in \mathbb{C}^n : \quad 2 \operatorname{Re} \left\{ \sum_{\alpha=1}^{n} w_\alpha \frac{\partial \psi}{\partial z_\alpha} (\xi_0) \right\} = 1 \right\} .$$

Consequently

$$\frac{\partial g}{\partial w_\alpha} (\xi_0, w) = -(n-1) \left\{ \frac{\bar{w}_\alpha}{\|w\|^{2n}} - \frac{\bar{w}_\alpha - N_\alpha}{\|w-\bar{N}\|^{2n}} \right\} \tag{11.2}$$

so that

$$\|\mathrm{Grad}_{(w)} g(\xi_o, w)\| = \frac{n-1}{\|\mathrm{Grad}\psi(\xi_0)\| \ \|w\|^{2n}} \quad \text{for} \quad w \in \partial D(\xi_0) . \tag{11.3}$$

We had introduced the functions

$$g_0(\xi, w) \equiv g(\xi, w) + \frac{1}{n-1} \sum_{\alpha=1}^{n} w_\alpha \frac{\partial g}{\partial w_\alpha} (\xi, w)$$

(11.4) and

$$g_\alpha(\xi, w) \equiv \psi(\xi) \frac{\partial g}{\partial \xi_\alpha} (\xi, w) - (n-1) \frac{\partial \psi}{\partial \xi_\alpha} (\xi)[g_0(\xi, w) + \bar{g}_0(\xi, w)]$$

for $\xi \in \bar{D}$ and $w \in D(\xi)$, $\alpha = 1, \dots, n$, in Chapter 6. Note that for $\xi \in D$ and $w \in D(\xi)$,

$$g_0(\xi, w) = G_0(\xi, z)\psi(\xi)^{2n-2} \quad \text{and} \quad g_\alpha(\xi, w) = G_\alpha(\xi, z)\psi(\xi)^{2n-1} .$$

In particular

$$g_0(\xi, 0) = G_0(\xi, \xi)\psi(\xi)^{2n-2} = \Lambda(\xi)\psi(\xi)^{2n-2} \equiv \lambda(\xi)$$

(11.5) and

$$g_\alpha(\xi, 0) = G_\alpha(\xi, \xi)\psi(\xi)^{2n-1} = \frac{\partial \Lambda}{\partial \xi_\alpha} (\xi) \ \psi(\xi)^{2n-1}, \quad \alpha = 1, \dots, n .$$

From explicit computation using (11.1) and (11.2), when $\xi = \xi_0$ formula (11.4) becomes

$$g_0(\xi_0, w) = \frac{1}{\|\mathrm{Grad}\psi(\xi_0)\|^2} \left[\frac{\sum_{\alpha=1}^{n} \bar{w}_\alpha \frac{\partial\psi}{\partial\bar{\xi}_\alpha}(\xi_0) - 1}{\|w - \bar{N}\|^{2n}} \right]$$

(11.6) and

$$g_\alpha(\xi_0, w) = -(n-1) \frac{\frac{\partial\psi}{\partial\xi_\alpha}(\xi_0)}{\|\mathrm{Grad}\psi(\xi_0)\|^2} \left[\frac{\sum_{\alpha=1}^{n} \left[\frac{\partial\psi}{\partial\xi_\alpha}(\xi_0) w_\alpha + \frac{\partial\psi}{\partial\bar{\xi}_\alpha}(\xi_0)\bar{w}_\alpha \right] - 2}{\|w - \bar{N}\|^{2n}} \right]$$

For future reference, we record

(11.7)

$$g_0(\xi_0, w) - \bar{g}_0(\xi_0, w) = \frac{1}{\|\mathrm{Grad}\psi(\xi_0)\|^2} \left[\frac{\sum_{\alpha=1}^{n} \left[\bar{w}_\alpha \frac{\partial\psi}{\partial\bar{\xi}_\alpha}(\xi_0) - w_\alpha \frac{\partial\psi}{\partial\xi_\alpha}(\xi_0) \right]}{\|w - \bar{N}\|^{2n}} \right] .$$

Proposition 11.1. Let $D = \{z \in \mathbb{C}^n : \psi(z) < 0\}$ be a bounded pseudoconvex domain in $\mathbb{C}^n$ with smooth boundary. Let $a \in S_n$ and $\xi_0 \in \partial D$. If

(11.8)
$$\sum_{\alpha=1}^{n} a_\alpha \frac{\partial\psi}{\partial\xi_\alpha}(\xi_0) \neq 0 ,$$

then

(11.9)
$$\varliminf_{\substack{\xi\to\xi_0 \\ \xi\in D}} \ell_2(\xi, a)/|\ell_1(\xi, a)|^2 \geq \frac{1}{2(n-1)} .$$

Remark. Condition (11.8) says that the vector $a = (a_1, \ldots, a_n)$ does not lie in the complex tangent space to ∂D at ξ_0.

Proof. Recall from Chapter 9, equations (9.5), (9.7) and (9.8), that

$$(9.7)\qquad \ell_2(\xi,a) \geq \tilde{J}(\xi,a) = \frac{4}{(n-1)w_{2n}(-\lambda(\xi))}\iint_{D(\xi)} \sum_{\alpha=1}^{n} \left|\frac{\partial}{\partial \bar{w}_\alpha} h(a,\xi,w)\right|^2 dV_w$$

where

$$(9.8)\qquad h(a,\xi,w) = -\ell_1(\xi,a)g_0(\xi,w) + \frac{1}{\psi(\xi)}\sum_{\alpha=1}^{n} a_\alpha g_\alpha(\xi,w)$$

$$= -\ell_1\left(\frac{g_0-\bar{g}_0}{2}\right) + \sum_{\alpha=1}^{n} a_\alpha \frac{\partial g}{\partial \xi_\alpha} - \left(\frac{g_0+\bar{g}_0}{2}\right)\left(\sum_{\alpha=1}^{n} a_\alpha \frac{\frac{\partial \lambda}{\partial \xi_\alpha}}{\lambda}\right)$$

for $w \in D(\xi)$. Writing

$$\ell_1(\xi,a) = \frac{\sum_{\alpha=1}^{n} a_\alpha \frac{\partial \Lambda}{\partial \xi_\alpha}(\xi)}{\Lambda(\xi)} = \frac{\sum_{\alpha=1}^{n} a_\alpha \frac{\partial \Lambda}{\partial \xi_\alpha}(\xi)\,\psi(\xi)^{2n-1}}{\Lambda(\xi)\psi(\xi)^{2n-2}\psi(\xi)},$$

it follows from Lemma 6.1 (2) and the hypothesis (11.8) that the numerator in the above equation tends to

$$(2n-2)\left[\sum_{\alpha=1}^{n} a_\alpha \frac{\partial \psi}{\partial \xi_\alpha}(\xi_0)\right] \|\mathrm{Grad}\psi(\xi_0)\|^{2n-2} = A \neq 0,\infty$$

as $\xi \to \xi_0$. From Lemma 6.1 (1) and the fact that $\psi(\xi_0) = 0$, the denominator tends to

$$-\|\mathrm{Grad}\psi(\xi_0)\|^{2n-2}\psi(\xi_0) = 0 .$$

Hence

$$\lim_{\xi\to\xi_0} \ell_1(\xi,a) = \infty .$$

On the other hand, since $\lambda(\xi_0) = -\|\mathrm{Grad}\psi(\xi_0)\|^{2n-2} \neq 0$,

$$\lim_{\xi\to\xi_0}\left[\sum_{\alpha=1}^{n} a_\alpha \frac{\partial g}{\partial \xi_\alpha}(\xi,w) - \frac{g_0(\xi,w)+\bar{g}_0(\xi,w)}{2}\left(\sum_{\alpha=1}^{n}\frac{a_\alpha \frac{\partial\lambda}{\partial\xi_\alpha}(\xi)}{\lambda(\xi)}\right)\right]$$

exists and is finite. Thus from (9.8),

$$\lim_{\xi\to\xi_0}\frac{h(a,\xi,w)}{\ell_1(\xi,a)} = -\left[\frac{g_0(\xi_0,w)-\overline{g_0}(\xi_0,w)}{2}\right]$$

for any $w \in D(\xi_0)$. Using (11.7) and Fatou's lemma we get the estimate

(11.10)

$$\varliminf_{\xi\to\xi_0}\frac{\ell_2(\xi,a)}{|\ell_1(\xi,a)|^2} \geq \varliminf_{\xi\to\xi_0}\frac{\tilde{J}(\xi,a)}{|\ell_1(\xi,a)|^2}$$

$$\geq \frac{4}{(n-1)w_{2n}(-\lambda(\xi_0))}\iint_{D(\xi_0)}\sum_{\alpha=1}^{n}\left|\frac{\partial}{\partial\bar{w}_\alpha}\left[\frac{g_0(\xi_0,w)-\overline{g_0}(\xi_0,w)}{2}\right]\right|^2 dV_w$$

$$= \frac{1}{(n-1)w_{2n}\|\mathrm{Grad}\psi(\xi_0)\|^{2n+2}}$$

$$\cdot\iint_{D(\xi_0)}\sum_{\alpha=1}^{n}\left|\frac{\partial}{\partial\bar{w}_\alpha}\left[\frac{\sum_{\beta=1}^{n}\left(\bar{w}_\beta\frac{\partial\psi}{\partial\bar{\xi}_\beta}(\xi_0) - w_\beta\frac{\partial\psi}{\partial\xi_\beta}(\xi_0)\right)}{\|w-\bar{N}\|^{2n}}\right]\right|^2 dV_w \equiv J^*(\xi_0).$$

We now compute $J^*(\xi_0)$. We can assume that $\mathrm{Grad}\psi(\xi_0) = (0,\ldots,0,i)$ so that the hyperplane $L = \partial D(\xi_0)$ is given by

$$L = \{w \in \mathbb{C}^n: \quad 2\,\mathrm{Re}(iw_n) = -2y_{2n} = 1\}$$

where $w = (w_1,\dots,w_n) = (y_1,y_2,\dots,y_{2n-1},y_{2n})$, i.e., $w_\alpha = y_{2\alpha-1}+iy_{2\alpha}$. Thus $D(\xi_0) = \{w \in \mathbb{C}^n: y_{2n} > -\frac{1}{2}\}$ and $\bar{N} = (0,\dots,0,-i)$. We set

$$u(w) \equiv \frac{\sum_{\beta=1}^{n}\left(\bar{w}_\beta \frac{\partial\psi}{\partial\bar{\xi}_\beta}(\xi_0) - w_\beta \frac{\partial\psi}{\partial\xi_\beta}(\xi_0)\right)}{\|w - \bar{N}\|^{2n}} = \frac{-\bar{w}_n i - w_n i}{[|w_1|^2+\dots+|w_n+i|^2]^n}$$

$$= \frac{-2iy_{2n-1}}{[y_1^2+\dots+y_{2n-1}^2+(y_{2n}+1)^2]^n}$$

Then $u(w)$ is a harmonic function in $D(\xi_0)$ and takes on purely imaginary values. We have

$$J^*(\xi_0) = \frac{1}{(n-1)w_{2n}} \iint_{D(\xi_0)} \sum_{\alpha=1}^{n} \left|\frac{\partial u}{\partial \bar{w}_\alpha}\right|^2 dV_w$$

$$= \frac{1}{4(n-1)w_{2n}} \int_{\partial D(\xi_0)} u \frac{\partial u}{\partial n_w}\, dS_w$$

$$= \frac{1}{(n-1)w_{2n}} \int_{-\infty}^{\infty} \cdots \int_{-\infty}^{\infty} \frac{y_{2n-1}}{(y_1^2+\dots+y_{2n-1}^2 + \frac{1}{4})^n}$$

$$\cdot \left[\frac{\partial}{\partial y_{2n}} \frac{y_{2n-1}}{[y_1^2+\dots+y_{2n-1}^2 + (y_{2n}+1)^2]^n}\right]\Bigg|_{y_{2n}=-\frac{1}{2}} dy_1\dots dy_{2n-1}$$

$$= \frac{n}{(n-1)w_{2n}} \int_{-\infty}^{\infty} \cdots \int_{-\infty}^{\infty} \frac{y_{2n-1}^2}{(y_1^2+\dots+y_{2n-1}^2 + \frac{1}{4})^{2n+1}}\, dy_1\dots dy_{2n-1}$$

To compute this integral, note that for $c > 0$ and $m > 1$,

$$\int_{-\infty}^{\infty} \frac{x^2}{(x^2+c)^{m+1}}\, dx = 2\int_0^{\infty} \frac{x^2}{(x^2+c)^{m+1}}\, dx = -\frac{1}{m}\frac{x}{(x^2+c)^m}\Big|_0^{\infty}$$

$$+\frac{1}{m}\int_0^{\infty}\frac{1}{(x^2+c)^m}\, dx = \frac{1}{2m}\int_{-\infty}^{\infty}\frac{1}{(x^2+c)^m}\, dx$$

Thus letting $m = 2n$, $c = y_1^2+\ldots+y_{2n-2}^2 + \frac{1}{4}$, and $x = y_{2n-1}$, we obtain

$$J^*(\xi_0) = \frac{1}{4(n-1)w_{2n}}\int_{-\infty}^{\infty}\cdots\int_{-\infty}^{\infty}\frac{1}{(y_1^2+\ldots+y_{2n-1}^2+\frac{1}{4})^{2n}}\, dy_1\ldots dy_{2n-1}$$

(11.11)

$$= \frac{1}{4(n-1)}\frac{w_{2n-1}}{w_{2n}}\int_0^{\infty}\frac{r^{2n-2}}{(r^2+\frac{1}{4})^{2n}}\, dr \equiv \frac{1}{4(n-1)}\frac{w_{2n-1}}{w_{2n}}\, I(2n-2,2n)$$

using polar coordinates. Repeated integration by parts yields

$$I(2n-2,2n) = \frac{2n-3}{2(2n-1)}\,\frac{2n-5}{2(2n-2)}\cdots\frac{1}{2(n+1)}\, I(0,n+1)$$

The residue theorem yields

$$I(0,n+1) \equiv \int_0^{\infty}\frac{1}{(r^2+\frac{1}{4})^{n+1}}\, dr = \frac{1}{2}\int_{-\infty}^{\infty}\frac{1}{(x^2+\frac{1}{4})^{n+1}}\, dx = \frac{\pi}{n!}(n+1)(n+2)\ldots(2n)$$

so that, using $w_m = 2\pi^{m/2}/\Gamma(\frac{m}{2})$,

(11.12)

$$J^*(\xi_0) = \frac{1}{4(n-1)}\left(\frac{1}{2}\right)^{n-1}\frac{w_{2n-1}}{w_{2n}}\left[\frac{1\cdot 3\ \ldots\ (2n-5)(2n-3)}{(n+1)(n+2)\ldots(2n-2)(2n-1)}\right]\frac{\pi}{n!}(n+1)(n+2)\ldots(2n)$$

$$= \frac{1}{2(n-1)}\ . \quad \blacksquare$$

Remark. Proposition 11.1 shows that, at least for non-complex tangential approach to ∂D, the relationship between the quantities $\ell_2(\xi, a)$ and $\ell_1(\xi, a)$ near ∂D behaves much like in the convex case (c.f., equation (9.4) in Chapter 9).

For an application in Chapter 12, we next study the behavior of the derivative $\frac{\partial\lambda}{\partial\xi_\alpha}$ of the function $\lambda(\xi)$ as $\xi \to \xi_0 \in \partial D$. We still take $\mathrm{Grad}\psi(\xi_0) = (0,\ldots,0,i)$ so that

$$g(\xi_0, w) = \frac{1}{\|w\|^{2n-2}} - \frac{1}{\|w-(0,\ldots,0,-i)\|^{2n-2}}$$

and from (11.6) and (11.2)

$$g_0(\xi_0, w) = \frac{-i\bar{w}_n - 1}{\|w-(0,\ldots,0,-i)\|^{2n-2}} \qquad \text{and}$$

$$\frac{\partial g}{\partial w_\alpha}(\xi_0, w) = -(n-1)\left[\frac{\bar{w}_\alpha}{\|w\|^{2n}} - \frac{\bar{w}_\alpha - \delta_{\alpha n} i}{\|w-(0,\ldots,0,-i)\|^{2n}}\right]$$

for $\alpha = 1,\ldots,n$. Also,

$$\|\mathrm{Grad}_{(w)} g(\xi_0, w)\| = \frac{n-1}{\|w\|^{2n}} \tag{11.13}$$

for $w \in L = \partial D(\xi_0)$ by (11.3).

Proposition 11.2. Under the above hypothesis,

$$\frac{\partial\lambda}{\partial\xi_\alpha}(\xi_0) = \begin{cases} (n-1)i\,[\psi_{\alpha n} - \psi_{\alpha\bar{n}}], & 1 \le \alpha \le n-1 \\ (n-1)i[\psi_{nn} - \psi_{n\bar{n}}] + \dfrac{i}{2(2n-1)}[\Delta\psi + (n-1)\psi_{n\bar{n}}], & \alpha = n \end{cases} \tag{11.14}$$

where $\psi_\alpha = \frac{\partial\psi}{\partial\xi_\alpha}(\xi_0)$, $\psi_{\bar{\alpha}} = \frac{\partial\psi}{\partial\bar{\xi}_\alpha}(\xi_0)$, $\psi_{\alpha\beta} = \frac{\partial^2\psi}{\partial\xi_\alpha\partial\xi_\beta}(\xi_0)$, etc.

Proof. Since $\frac{\partial\lambda}{\partial\xi_\alpha}(\xi) = \frac{\partial g}{\partial\xi_\alpha}(\xi,0)$, we recall from Corollary 4.1 that

$$\frac{\partial g}{\partial\xi_\alpha}(\xi,w) = \frac{1}{2(n-1)w_{2n}}\int_{\partial D(\xi)} K_1(\xi,\eta)\|\mathrm{Grad}_{(\eta)}g(\xi,\eta)\|\ \frac{\partial g_w(\xi,\eta)}{\partial n_\eta}\,dS_\eta$$

where $g_w(\xi,\eta)$ is the Green function for $D(\xi)$ with pole at $\eta = w$ and $K_1(\xi,\eta) = K_1^{(\alpha)}(\xi,\eta) = \frac{\partial f}{\partial\xi_\alpha}/\|\mathrm{Grad}_{(\eta)}f(\xi,\eta)\|$, i.e., for $\xi \in D_1$, $\frac{\partial g}{\partial\xi_\alpha}(\xi,w)$ is a harmonic function of $w \in D(\xi)$ with boundary values

$$-K_1(\xi,\eta)\|\mathrm{Grad}_{(\eta)}g(\xi,\eta)\|,\quad \eta \in \partial D(\xi)\ .$$

We want to analyze this quantity as $\xi \to \xi_0 \in \partial D$. Recall, first of all, that

(11.15) $$\|\mathrm{Grad}_{(\eta)}f(\xi,\eta)\| = \|\mathrm{Grad}\ \psi(\xi-\psi(\xi)\eta)\|$$

since $\frac{\partial f}{\partial w_\alpha}(\xi,w) = \frac{\partial\psi}{\partial z_\alpha}(\xi - \psi(\xi)w)$ where $w = \frac{z-\xi}{-\psi(\xi)}$. Evaluating (11.15) at ξ_0, we obtain that

(11.16) $$\|\mathrm{Grad}_{(\eta)}f(\xi_0,\eta)\| = \|\mathrm{Grad}\ \psi(\xi_0)\| = 1\ .$$

From Equation (4.14) we have, using (11.16) and the definition of k_1,

(11.17)

$$\begin{aligned}\lim_{\xi\to\xi_0} K_1(\xi,\eta) &= K_1(\xi_0,\eta) = \frac{\partial f}{\partial\xi_\alpha}(\xi_0,\eta)\\ &= \sum_{\beta=1}^{n}\left[\eta_\beta\frac{\partial^2\psi}{\partial\xi_\alpha\partial\xi_\beta}(\xi_0) + \bar\eta_\beta\frac{\partial^2\psi}{\partial\xi_\alpha\partial\bar\xi_\beta}(\xi_0)\right]\\ &\quad - \frac{i}{2}\delta_{\alpha n}\sum_{\beta,\gamma=1}^{n}\left[\eta_\beta\eta_\gamma\frac{\partial^2\psi}{\partial\xi_\beta\partial\xi_\gamma}(\xi_0) + \bar\eta_\beta\bar\eta_\gamma\frac{\partial^2\psi}{\partial\bar\xi_\beta\partial\bar\xi_\gamma}(\xi_0)\right.\\ &\qquad\left. + \eta_\beta\bar\eta_\gamma\frac{\partial^2\psi}{\partial\xi_\beta\partial\bar\xi_\gamma}(\xi_0) + \bar\eta_\beta\eta_\gamma\frac{\partial^2\psi}{\partial\bar\xi_\beta\partial\xi_\gamma}(\xi_0)\right]\end{aligned}$$

Since $K_1(\xi_0,\eta) = O(\|\eta\|^2)$ and $\|\mathrm{Grad}_{(\eta)} g(\xi_0,\eta)\| = O\left(\frac{1}{\|\eta\|^{2n}}\right)$, the integral

$$\int_{\partial D(\xi_0)} K_1(\xi_0,\eta)\|\mathrm{Grad}_{(\eta)} g(\xi_0,\eta)\| \frac{\partial g_w(\xi_0,\eta)}{\partial n_\eta} dS_\eta$$

converges so that, using (11.13),

(11.18)

$$\frac{\partial g}{\partial \xi_\alpha}(\xi_0,w) = \lim_{\xi\to\xi_0} \frac{\partial g}{\partial \xi_\alpha}(\xi,w) = \frac{1}{2(n-1)w_{2n}} \int_{\partial D(\xi_0)} K_1(\xi_0,\eta) \frac{n-1}{\|\eta\|^{2n}} \frac{\partial g_w(\xi_0,\eta)}{\partial n_\eta} dS_\eta$$

$$\equiv -(n-1)\, H_{K_1(\xi_0,\eta)/\|\eta\|^{2n}}$$

(recall the notation introduced in Chapter 5: $H_{P(\eta)}$ = harmonic function in $D(\xi_0)$ with boundary values $P(\eta)$) . To simplify the notation in what follows, we let

$$h_\beta(w) = H_{\eta_\beta/\|\eta\|^{2n}}\ ;\ h_{\beta\gamma}(w) = H_{\eta_\beta\eta_\gamma/\|\eta\|^{2n}};\ h_{\beta\bar\gamma}(w) = H_{\eta_\beta\bar\eta_\gamma/\|\eta\|^{2n}}\ .$$

and equation (11.18) becomes, using (11.17),

(11.18)′

$$\frac{\partial g}{\partial \xi_\alpha}(\xi_0,w) = -(n-1)\left[\sum_{\beta=1}^{n}\left[\psi_{\alpha\beta}h_\beta(w) + \psi_{\alpha\bar\beta}\overline{h_\beta(w)}\right]\right.$$

$$\left. - \frac{i}{2}\delta_{\alpha n}\sum_{\beta,\gamma=1}^{n}\left[\psi_{\beta\gamma}h_{\beta\gamma}(w) + \psi_{\bar\beta\bar\gamma}\overline{h_{\beta\gamma}(w)} + \psi_{\beta\bar\gamma}h_{\beta\bar\gamma}(w) + \psi_{\bar\beta\gamma}\overline{h_{\beta\bar\gamma}(w)}\right]\right] .$$

Our next task is to compute the functions $h_\beta(w)$, $h_{\beta\gamma}(w)$ and $h_{\beta\bar\gamma}(w)$ at $w = 0$.

To compute $h_\beta(w)$, $\beta = 1,\ldots,n$, we note first that

$$P(w) \equiv \frac{1}{\|w-(0,\dots,0,-i)\|^{2n-2}}$$

is harmonic for $w \in D(\xi_0)$ with boundary values $P(\eta) = \dfrac{1}{\|\eta\|^{2n-2}}$, $\eta \in \partial D(\xi_0)$. Since a derivative of a harmonic function is harmonic and

$$\frac{\partial}{\partial \bar{w}_\beta} P(w) = -(n-1) \frac{w_\beta}{\|w-(0,\dots,0,-i)\|^{2n}}, \quad \beta = 1,\dots,n-1,$$

we see that

$$h_\beta(w) = \frac{w_\beta}{\|w-(0,\dots,0,-i)\|^{2n}}, \quad \beta = 1,\dots,n-1.$$

Similarly,

$$h_n(w) = \frac{-1}{(n-1)} \frac{\partial}{\partial w_n} P(w) = \frac{\overline{w_n + i}}{\|w-(0,\dots,0,-i)\|^{2n}}$$

since $\eta_n = y_{2n-1} - \frac{1}{2} i$. In particular,

(11.19) $$h_\beta(0) = 0 \quad \text{for} \quad \beta = 1,\dots,n-1 \quad \text{while} \quad h_n(0) = -i.$$

For future use, we note that $g_0(\xi_0, w) = -ih_n(w)$ from (11.6), hence

(11.20) $$g_0(\xi_0, w) - \overline{g_0(\xi_0, w)} = -i[h_n(w) + \overline{h_n(w)}] \quad \text{and}$$

$$g_0(\xi_0, w) + \overline{g_0(\xi_0, w)} = -i[h_n(w) - \overline{h_n(w)}].$$

To compute $h_{\alpha\beta}(w)$, since $g_w(\xi_0, \eta) = \dfrac{1}{\|\eta-w\|^{2n-2}} - \dfrac{1}{\|\eta-w^*\|^{2n-2}}$, we have

$$h_{\alpha\beta}(w) = \frac{1}{2(n-1)w_{2n}} \int_{\partial D(\xi_0)} \frac{\eta_\alpha \eta_\beta}{\|\eta\|^{2n}} \frac{\partial}{\partial n_\eta} \left[\frac{1}{\|\eta-w\|^{2n-2}} - \frac{1}{\|\eta-w^*\|^{2n-2}} \right] dS_\eta$$

where $w^* = (w_1,\dots,w_{n-1},\bar{w}_n - i)$ is the symmetric point of w with respect to $L = \partial D(\xi_0)$. Writing $\eta_\alpha = y_{2\alpha-1} + iy_{2\alpha}$ and $w_\alpha = x_{2\alpha-1} + ix_{2\alpha}$, $\alpha = 1,\ldots,n$, we have

(11.21)

$$h_{\alpha\beta}(w) = \frac{-1}{2(n-1)w_{2n}} \int_{-\infty}^{\infty} \cdots \int_{-\infty}^{\infty} \frac{(y_{2\alpha-1}+iy_{2\alpha})(y_{2\beta-1}+iy_{2\beta})}{(y_1^2 + \ldots + y_{2n-1}^2 + \frac{1}{4})^n}$$

$$\cdot \frac{\partial}{\partial y_{2n}}\left[\frac{1}{[(y_1-x_1)^2 + \ldots + (y_{2n}-x_{2n})^2]^{n-1}} \right.$$

$$\left. - \frac{1}{[(y_1-x_1)^2 + \ldots + (y_{2n}+x_{2n}+1)^2]^{n-1}} \right] \Bigg|_{y_{2n} = -\frac{1}{2}} dy_1 \ldots dy_{2n-1}$$

$$\equiv \frac{2x_{2n}+1}{w_{2n}} \left[I_{2\alpha-1,2\beta-1}(x) - I_{2\alpha,2\beta}(x) + i(I_{2\alpha-1,2\beta}(x) + I_{2\alpha,2\beta-1}(x)) \right]$$

where

$I_{k\ell}(x) \equiv$

$$\int_{-\infty}^{\infty} \cdots \int_{-\infty}^{\infty} \frac{y_k y_\ell}{[y_1^2 + \ldots + y_{2n-1}^2 + \frac{1}{4}]^n [(y_1-x_1)^2 + \ldots + (y_{2n-1}-x_{2n-1})^2 + (x_{2n} + \frac{1}{2})^2]^n}$$

$$dy_1 \ldots dy_{2n-1}$$

We can get analogous formulas for $h_{\alpha\bar{\beta}}(w)$ and $h_{\bar{\alpha}\beta}(w)$. When $\alpha = \beta = n$, we get a relationship between $h_{n\bar{n}}(w)$, $h_{\bar{n}n}(w)$, $h_n(w)$ and $h_{\bar{n}}(w)$. Since $\eta \in \partial D(\xi_0)$ implies that $\eta_n = y_{2n-1} - \frac{1}{2} i$,

$$\bar{\eta}_n^2 - \eta_n^2 = 2iy_{2n-1} = i(\bar{\eta}_n + \eta_n) .$$

It follows that

$$h_{\bar{n}\bar{n}}(w) - h_{nn}(w) = i[h_{\bar{n}}(w)+h_{n}(w)]$$

In particular, setting $w = 0$ and using (11.19), we obtain

(11.22) $$h_{\bar{n}\bar{n}}(0) = h_{nn}(0) .$$

Before computing $h_{\alpha\beta}(0)$ and $h_{\alpha\bar{\beta}}(0)$, $\alpha,\beta = 1,\ldots,n$, we introduce some notation. Let

$$Y \equiv y_1^2 + \ldots + y_{2n-1}^2 + \frac{1}{4} .$$

Then, for $1 \le \alpha,\beta \le n-1$,

$$h_{\alpha\beta}(0) = \frac{1}{2(n-1)w_{2n}} \int_{\partial D(\xi_0)} \frac{\eta_\alpha \eta_\beta}{\|\eta\|^{2n}} \frac{\partial g(\xi_0,\eta)}{\partial n_\eta} dS_\eta$$

$$= \frac{1}{2(n-1)w_{2n}} \int_{-\infty}^{\infty} \cdots \int_{-\infty}^{\infty} \frac{(y_{2\alpha-1}+iy_{2\alpha})(y_{2\beta-1}+iy_{2\beta})}{Y^{2n}} dy_1\ldots dy_{2n-1}$$

By symmetry,

(11.23) $$h_{\alpha\beta}(0) = 0, \quad 1 \le \alpha,\beta \le n-1 .$$

Similarly,

(11.24)

$$h_{\alpha\bar{\beta}}(0) = \frac{1}{2(n-1)w_{2n}} \int_{-\infty}^{\infty} \cdots \int_{-\infty}^{\infty} \frac{(y_{2\alpha-1}+iy_{2\alpha})(y_{2\beta-1}-iy_{2\beta})}{Y^{2n}} dy_1\ldots dy_{2n-1}$$

$$= 0 \quad \text{if} \quad 1 \le \alpha,\beta \le n-1 \quad \text{and} \quad \alpha \ne \beta .$$

If $\alpha = \beta \le n-1$, then

$$h_{\beta\bar{\beta}}(0) = \frac{1}{2(n-1)w_{2n}} \int_{-\infty}^{\infty} \cdots \int_{-\infty}^{\infty} \frac{y_{2\beta-1}^2 + y_{2\beta}^2}{Y^{2n}} \, dy_1 \ldots dy_{2n-1}$$

$$= \frac{1}{(n-1)w_{2n}} \int_{-\infty}^{\infty} \cdots \int_{-\infty}^{\infty} \frac{y_1^2}{Y^{2n}} \, dy_1 \ldots dy_{2n-1} \, .$$

Integrating by parts in y_1, we obtain

$$\int_{-\infty}^{\infty} \cdots \int_{-\infty}^{\infty} \frac{y_1^2}{Y^{2n}} \, dy_1 \ldots dy_{2n-1} = \frac{1}{2(2n-1)} \int_{-\infty}^{\infty} \cdots \int_{-\infty}^{\infty} \frac{1}{Y^{2n-1}} \, dy_1 \ldots dy_{2n-1}$$

(11.25)
$$= \frac{w_{2n-1}}{2(2n-1)} \int_0^{\infty} \frac{r^{2n-2}}{(r^2 + \frac{1}{4})^{2n-1}} \, dr$$

$$\equiv \frac{w_{2n-1}}{2(2n-1)} \, I(2n-2, 2n-1) \, .$$

In a manner analogous to the computation of $I(2n-2, 2n)$ in (11.11), we find

$$I(2n-2, 2n-1) = \pi \, \frac{1 \cdot 3 \ldots (2n-3)}{2 \cdot 4 \ldots (2n-2)} \, .$$

Hence

(11.26)
$$h_{\beta\bar{\beta}}(0) = \frac{1}{(n-1)w_{2n}} \, \frac{w_{2n-1}}{2(2n-1)} \, \pi \, \frac{1 \cdot 3 \ldots (2n-3)}{2 \cdot 4 \ldots (2n-2)} = \frac{1}{2(n-1)(2n-1)}$$

for $\beta = 1, \ldots, n-1$.

Finally, when $w = 0$ and $\alpha = n$ in (11.21), we obtain

(11.27)

$$h_{n\beta}(0) = \frac{1}{2(n-1)w_{2n}} \int_{-\infty}^{\infty} \cdots \int_{-\infty}^{\infty} \frac{(y_{2\beta-1} + iy_{2\beta})(y_{2n-1} - \frac{1}{2}i)}{Y^{2n}} \, dy_1 \ldots dy_{2n-1} = 0,$$

$\beta = 1, \ldots, n-1$

by symmetry; similarly,

(11.28) $$h_{n\bar{\beta}}(0) = 0\ ,\ \beta = 1,\ldots,n-1\ .$$

Last of all, to compute $h_{n\bar{n}}(0) = \frac{1}{2(n-1)w_{2n}}\int_{-\infty}^{\infty}\cdots\int_{-\infty}^{\infty}\frac{y_{2n-1}^2+\frac{1}{4}}{Y^{2n}}\,dy_1\ldots dy_{2n-1}$,

note that

$$\frac{1}{Y^{2n-1}} = \frac{Y}{Y^{2n}} = \frac{y_1^2+\ldots+\frac{1}{4}}{Y^{2n}}$$

implies that

$$\int_{-\infty}^{\infty}\cdots\int_{-\infty}^{\infty}\frac{1}{Y^{2n-1}}\,dy_1\ldots dy_{2n-1} = (2n-1)\int_{-\infty}^{\infty}\cdots\int_{-\infty}^{\infty}\frac{y_1^2}{Y^{2n}}\,dy_1\ldots dy_{2n-1}$$

$$+\frac{1}{4}\int_{-\infty}^{\infty}\cdots\int_{-\infty}^{\infty}\frac{1}{Y^{2n}}\,dy_1\ldots dy_{2n-1}$$

$$=\frac{1}{2}\int_{-\infty}^{\infty}\cdots\int_{-\infty}^{\infty}\frac{1}{Y^{2n-1}}\,dy_1\ldots dy_{2n-1}$$

$$+\frac{1}{4}\int_{-\infty}^{\infty}\cdots\int_{-\infty}^{\infty}\frac{1}{Y^{2n}}\,dy_1\ldots dy_{2n-1}$$

by (11.25). Thus

$$\int_{-\infty}^{\infty}\cdots\int_{-\infty}^{\infty}\frac{1}{Y^{2n}}\,dy_1\ldots dy_{2n-1} = 2\int_{-\infty}^{\infty}\cdots\int_{-\infty}^{\infty}\frac{1}{Y^{2n-1}}\,dy_1\ldots dy_{2n-1}\ ;$$

thus, again using (11.25), we see that

(11.29)

$$h_{n\bar{n}}(0) = \left[\frac{1}{2(2n-1)}+\frac{1}{2}\right]\frac{1}{2(n-1)w_{2n}}\int_{-\infty}^{\infty}\cdots\int_{-\infty}^{\infty}\frac{1}{Y^{2n-1}}\,dy_1\ldots dy_{2n-1}$$

$$= \frac{n}{2(2n-1)(n-1)}\,\frac{w_{2n-1}}{w_{2n}}\,I(2n-2,2n-1) = \frac{n}{2(n-1)(2n-1)}\ .$$

We conclude, using (11.19), (11.22)-(11.24), and (11.26)-(11.29) in (11.18)′ when $w = 0$, that

(11.14)

$$\frac{\partial\lambda}{\partial\xi_\alpha}(\xi_0) = \begin{cases} (n-1)i[\psi_{\alpha n}-\psi_{\alpha\bar{n}}], & 1 \le \alpha \le n-1 \\ (n-1)i[\psi_{nn}-\psi_{n\bar{n}}] + \frac{i}{2(2n-1)}[\Delta\psi + (n-1)\psi_{n\bar{n}}], & \alpha = n \end{cases} . \blacksquare$$

As mentioned, this formula will be used in the next chapter, as will the following lemma.

Lemma 11.1. Let $\mathcal{E}$ denote the collection of functions

$$\mathcal{E} \equiv \{h_\gamma(w), h_{\bar{\gamma}}(w), h_{\alpha\beta}(w), h_{\bar{\alpha}\bar{\beta}}(w), h_{\alpha\bar{\beta}}(w), h_{\bar{\alpha}\beta}(w), h_{\gamma n}(w), h_{\bar{\gamma}n}(w)\}$$

where $\gamma = 1,\ldots,n$ and $1 \le \alpha,\beta \le n-1$. Then $\mathcal{E}$ is a linearly independent (over $\mathbb{C}$) collection of harmonic functions in $D(\xi_0)$ and

(1) if $f(w)$ is a holomorphic function in $D(\xi_0)$ which is in the linear span of $\mathcal{E}$, then $f(w) \equiv 0$ in $D(\xi_0)$.

Furthermore, the functions $h_{\gamma\bar{n}}, h_{\bar{\gamma}\bar{n}}$ satisfy

(2) $h_{\gamma\bar{n}}(w) = h_{\gamma n}(w) + ih_\gamma(w)$; $h_{\bar{\gamma}\bar{n}}(w) = h_{\bar{\gamma}n}(w) - ih_{\bar{\gamma}}(w)$, $\gamma = 1,\ldots,n-1$;

$h_{\bar{n}\bar{n}}(w) = h_{nn}(w) + i[h_n(w) + h_{\bar{n}}(w)]$;

$h_{n\bar{n}}(w) = h_{nn}(w) + ih_n(w)$.

Proof. The linear independence of $\mathcal{E}$ is clear by considering the boundary values of an element in $f(w)$ in the linear span of $\mathcal{E}$. To prove (1), note that, for such an $f(w)$, the boundary values $F(\eta)$, $\eta \in \partial D(\xi_0)$, satisfy

$$F(\eta) = \frac{1}{\left[|\eta_1|^2+\ldots+|\eta_{n-1}|^2+(\mathrm{Re}\,\eta_n)^2 + \frac14\right]^n} \cdot P_2(\eta_1,\ldots,\eta_{n-1},\mathrm{Re}\,\eta_n)$$

where $P_2(t_1,\dots,t_n)$ is a quadratic polynomial. Thus if $w_n^0 \in \mathbb{C}$ with $\operatorname{Im} w_n^0 > -\frac{1}{2}$, then clearly

$$(11.30) \qquad \lim_{(w_1,\dots,w_{n-1})\to\infty} f(w_1,\dots,w_{n-1},w_n^0) = 0 .$$

However, if f is holomorphic in $D(\xi_0)$, then $f(w_1,\dots,w_{n-1},w_n^0)$ is holomorphic in $(w_1,\dots,w_{n-1}) \in \mathbb{C}^{n-1}$. It then follows from (11.30) that $f(w_1,\dots,w_{n-1},w_n^0) \equiv 0$ in $\mathbb{C}^{n-1}$ so that $f \equiv 0$ in $D(\xi_0)$. The equations in (2) follow by considering boundary values of the corresponding harmonic functions. ∎

12. SUFFICIENT CONDITIONS FOR COMPLETENESS OF THE Λ-METRIC

We now utilize the calculations from the last chapter to give a slightly more general sufficient condition than strict pseudoconvexity in order for the Λ-metric associated to a smoothly bounded domain D to be complete. In order to formulate this condition, we need a preliminary definition.

Let $D = \{z \in \mathbb{C}^n: \psi(z) < 0\}$ be a bounded pseudoconvex domain with smooth boundary. Given $\xi_0 \in \partial D$, the complex tangent space to ∂D at ξ_0 is given by

$$\Pi = \left\{z \in \mathbb{C}^n: \sum_{\alpha=1}^{n} \frac{\partial\psi}{\partial\xi_\alpha}(\xi_0)\left(z_\alpha - (\xi_0)_\alpha\right) = 0\right\} \tag{12.1}$$

We parameterize Π by finding an $n \times (n-1)$ matrix A with complex entries satisfying

$$\operatorname{Grad}\psi(\xi_0) \circ A = \left(\frac{\partial\psi}{\partial\xi_1}(\xi_0), \dots, \frac{\partial\psi}{\partial\xi_n}(\xi_0)\right) \circ A = (0, \dots, 0) \in \mathbb{C}^{n-1}.$$

Then $z \in \Pi$ if and only if

$$z = \begin{pmatrix} z_1 \\ \vdots \\ z_n \end{pmatrix} = \xi_0 + Au \equiv \begin{pmatrix} (\xi_0)_1 \\ \vdots \\ (\xi_0)_n \end{pmatrix} + \begin{pmatrix} a_{11} & \cdots & a_{1,n-1} \\ \vdots & & \\ a_{n1} & \cdots & a_{n,n-1} \end{pmatrix}\begin{pmatrix} u_1 \\ \vdots \\ u_{n-1} \end{pmatrix} \tag{12.2}$$

where $u \in \mathbb{C}^{n-1}$. If we now restrict the defining function ψ to a neighborhood of ξ_0 intersected with π, then

$$\Phi(u) \equiv \psi(\xi_0 + Au) \tag{12.3}$$

is defined in a neighborhood of $u = 0$ in $\mathbb{C}^{n-1}$. If ∂D is pseudoconvex at ξ_0, then the matrix

$$\Phi_{j\bar{k}} \equiv \left[\frac{\partial^2 \Phi}{\partial u_j \partial \bar{u}_k} (0) \right]_{j,k=1,\ldots,n-1}$$

is positive semi-definite while if ∂D is strictly pseudoconvex at ξ_0, this matrix is positive definite, and, in particular, nonsingular. Furthermore, these conditions are invariant under biholomorphic coordinate changes. On the other hand, if we consider the matrix

$$\Phi_{jk} \equiv \left[\frac{\partial^2 \Phi}{\partial u_j \partial u_k} (0) \right]_{j,k=1,\ldots,n-1} ,$$

the invertibility of this matrix is <u>not</u> invariant under biholomorphisms; indeed, one can always find a biholomorphic change of coordinates so that $\frac{\partial^2 \Phi}{\partial u_j \partial u_k} (0) = 0$, $j,k = 1,\ldots,n-1$, in the new coordinates. However, simple calculations show that the invertibility of Φ_{jk} is, first of all, independent of the defining function ψ and, secondly, the invertibility remains invariant under unitary coordinate transformations of the parameter $u \in \mathbb{C}^{n-1}$ (note that we can normalize A in (12.2) so that $A(\bar{A}^t) = I_n =$ $n \times n$ identity matrix). We thus make the following definition.

<u>Definition 12.1</u>. A point $\xi_0 \in \partial D$ is called a <u>nondegenerate boundary point</u> if

$$\text{(12.4)} \qquad \det \Phi_{jk} = \det \left[\frac{\partial^2 \Phi}{\partial u_j \partial u_k} (0) \right]_{j,k=1,\ldots,n-1} \neq 0 .$$

If we assume that $\mathrm{Grad}\psi(\xi_0) = (0,\ldots,a)$ for some $a \in \mathbb{C} - \{0\}$, then

$$\Pi = \{(z_1,\ldots,z_n) \in \mathbb{C}^n: \ z_n = 0\}$$

and we can take

$$A = \begin{pmatrix} 1 & & \\ & \ddots & 1 \\ 0 & \cdots & 0 \end{pmatrix} = \begin{pmatrix} I_{n-1} \\ 0 \cdots 0 \end{pmatrix}, \quad \text{i.e.,} \quad (u_1,\ldots,u_{n-1}) = (z_1,\ldots,z_{n-1}) .$$

In this setting, the condition of pseudoconvexity at $\xi_0 \in \partial D$ implies that

$$\det\left[\frac{\partial^2\psi}{\partial z_\alpha \partial \bar{z}_\beta} (\xi_0) \right]_{\alpha,\beta=1,\ldots,n-1} \geq 0 .$$

If ξ_0 is a point of strict pseudoconvexity, then

$$\det\left[\frac{\partial^2\psi}{\partial z_\alpha \partial \bar{z}_\beta} (\xi_0) \right]_{\alpha,\beta=1,\ldots,n-1} > 0 , \tag{12.5}$$

while if ξ_0 is a nondegenerate boundary point, then

$$\det\left[\frac{\partial^2\psi}{\partial z_\alpha \partial z_\beta} (\xi_0) \right]_{\alpha,\beta=1,\ldots,n-1} \neq 0 . \tag{12.6}$$

Example. The domain $D = \{(z_1,z_2) \in \mathbb{C}^2: \ \psi(z_1,z_2) = 2\,\mathrm{Re}(z_1^2+z_2) + |z_2|^2 < 0\}$ is pseudoconvex; $(0,0)$ is a nondegenerate boundary point but is not a point of strict pseudoconvexity since

$$\frac{\partial^2\psi}{\partial z_1 \partial \bar{z}_1} (0,0) = 0 \quad \text{while} \quad \frac{\partial^2\psi}{\partial z_1 \partial z_1} (0,0) = 2 .$$

With these preliminaries, we can now state the main result of this chapter.

Theorem 12.1. Let D be a bounded pseudoconvex domain in $\mathbb{C}^n$ with smooth boundary. If each boundary point of D is either nondegenerate or strictly pseudoconvex, then

1. $m \equiv \inf\{\ell_2(\xi,a): \ (\xi,a) \in D \times S_n\} > 0$

2. there exists $c > 0$ such that

$$\ell_2(\xi, a) \geq c|\ell_1(\xi, a)|^2$$

for all $(\xi, a) \in D \times S_n$.

3. the Λ-metric is complete for D .

Remark. If $D \subset \mathbb{C}^2$, $(0,0) \in \partial D$, and $\frac{\partial\psi}{\partial z_1}(0,0) = 0$ (so that $\pi = \{(z_1, z_2): \ z_2 = 0\})$, then $(0,0)$ is neither nondegenerate nor strictly pseudoconvex precisely when

$$\psi(z_1, 0) = O(|z_1|^3)$$

Proof of Theorem 12.1. 1. We suppose that $m = 0$. In order to derive a contradiction, we start by noting that since $\log(-\Lambda)$ is strictly plurisubharmonic in D (Corollary 7.1), on any relatively compact subdomain $D' \subset\subset D$, we have

$$m' \equiv \inf\{\ell_2(\xi, a): \ (\xi, a) \in D' \times S_n\} > 0 .$$

Thus if $m = 0$, we can clearly choose a sequence $\{(\xi^j, a^j)\} \subset D \times S_n$ such that $\lim_{j\to+\infty} \xi^j = \xi_0 \in \partial D$, $\lim_{j\to+\infty} a^j = a \in S_n$, and

$$\lim_{j\to+\infty} \ell_2(\xi^j, a^j) = 0 . \tag{12.7}$$

We want to use Lemma 11.1 to show that the quantity

$$h(a, \xi_0, w) = -\ell_1(\xi_0, a)\left[\frac{g_0(\xi_0, w)-\overline{g_0}(\xi_0, w)}{2}\right] + \sum_{\beta=1}^{n} a_\beta \frac{\partial g}{\partial \xi_\beta}(\xi_0, w)$$
$$- \left[\frac{g_0(\xi_0, w)+\overline{g_0}(\xi_0, w)}{2}\right] \sum_{\beta=1}^{n} a_\beta \frac{\frac{\partial\lambda}{\partial\xi_\beta}(\xi_0)}{\lambda(\xi_0)} \tag{12.8}$$

is identically 0 for $w \in D(\xi_0)$.

By Proposition 6.2 and Fatou's lemma, it follows that

$$\underline{\lim}_{j\to+\infty} \iint_{D(\xi^j)} \left(\sum_{\alpha=1}^{n} \left| \frac{\partial}{\partial \bar{w}_\alpha} h(a^j, \xi^j, w) \right|^2 \right) dV_w \geq \iint_{D(\xi_0)} \left(\sum_{\alpha=1}^{n} \left| \frac{\partial}{\partial \bar{w}_\alpha} h(a, \xi_0, w) \right|^2 \right) dV_w$$

Since

$$\ell_2(\xi^j, a^j) \geq J(\xi^j, a^j) = \frac{4}{(n-1)w_{2n}(-\lambda(\xi^j))} \iint_{D(\xi^j)} \left(\sum_{\alpha=1}^{n} \left| \frac{\partial}{\partial \bar{w}_\alpha} h(a^j, \xi^j, w) \right|^2 \right) dV_w$$

and

$$\lim_{j\to+\infty} \lambda(\xi^j) = \lambda(\xi_0) = -\|\mathrm{Grad}\psi(\xi_0)\|^{2n-2} < 0$$

where $D = \{z \in \mathbb{C}^n : \psi(z) < 0\}$, it follows from (12.7) that

$$\iint_{D(\xi_0)} \left(\sum_{\alpha=1}^{n} \left| \frac{\partial}{\partial \bar{w}_\alpha} h(a, \xi_0, w) \right|^2 \right) dV_w = 0$$

Hence $h(a, \xi_0, w)$ is holomorphic for $w \in D(\xi_0)$. We now show that $h(a, \xi_0, w)$ belongs to the linear span of $\mathcal{E}$, defined in the last chapter; then Lemma 11.1 implies that, indeed,

(12.9) $$h(a, \xi_0, w) \equiv 0 \quad \text{for} \quad w \in D(\xi_0) .$$

We may assume that Grad $\psi(\xi_0) = (0, \ldots, 0, i)$. From Proposition 6.2 and the computations of the last chapter,

(i) $\lim_{j\to+\infty} \lambda(\xi^j) = \lambda(\xi_0) = -\|\mathrm{Grad}\ \psi(\xi_0)\|^{2n-2} = -1$;

(ii) $\lim_{j\to+\infty} [g_0(\xi^j, w) - \bar{g}_0(\xi^j, w)] = -i[h_n(w) + \overline{h_n(w)}]$

and

$$\text{(iii)}\quad \lim_{j\to+\infty} [g_0(\xi^j,w) + \bar{g}_0(\xi^j,w)] = -i[h_n(w)-\overline{h_n(w)}]$$

from (11.20);

$$\text{(iv)}\quad \lim_{j\to+\infty} \sum_{\beta=1}^{n} a_\beta \frac{\partial g}{\partial \xi_\beta}(\xi^j,w) = \sum_{\beta=1}^{n} a_\beta \frac{\partial g}{\partial \xi_\beta}(\xi_0,w)$$

$$= -(n-1)\left[\sum_{\beta=1}^{n} a_\beta \left(\sum_{\gamma=1}^{n}\left[\psi_{\beta\gamma}h_\gamma(w) + \psi_{\beta\bar{\gamma}}\overline{h_\gamma(w)}\right]\right)\right]$$

$$+ \frac{i(n-1)}{2} a_n \left(\sum_{\beta,\gamma=1}^{n}\left[\psi_{\beta\gamma}h_{\beta\gamma}(w) + \psi_{\bar{\beta}\bar{\gamma}}\overline{h_{\beta\gamma}(w)} + \psi_{\beta\bar{\gamma}}h_{\beta\bar{\gamma}}(w)\right.\right.$$

$$\left.\left. + \psi_{\bar{\beta}\gamma}h_{\bar{\beta}\gamma}(w)\right]\right)$$

from (11.18)'; and finally

$$\text{(v)}\quad \lim_{j\to+\infty} \sum_{\beta=1}^{n} a_\beta \frac{\frac{\partial\lambda}{\partial\xi_\beta}(\xi^j)}{\lambda(\xi^j)} = -\sum_{\beta=1}^{n} a_\beta \frac{\partial\lambda}{\partial\xi_\beta}(\xi_0)$$

$$= -(n-1)\, i \sum_{\beta=1}^{n} a_\beta (\psi_{\beta n}-\psi_{\beta\bar{n}}) - a_n \frac{i}{2(2n-1)}[\Delta\psi + (n-1)\psi_{n\bar{n}}]$$

from (i) and (11.14). From Theorem 9.2, there exist constants $c,d > 0$ such that

$$|\ell_1(\xi^j,a^j)| \le \frac{\sqrt{\ell_2(\xi^j,a^j)} + d}{c}, \quad j = 1,2, \ldots$$

Thus by taking a subsequence which we again call $\{(\xi^j,a^j)\}$

$$(12.10)\qquad \lim_{j\to+\infty} |\ell_1(\xi^j,a^j)| \equiv \ell^*$$

where $|\ell^*| \le d/c$. Using i) - v) and (12.10) and the definition of $h(a,\xi_0,w)$ in (12.8), we find that

(12.11)

$$\lim_{j\to+\infty} h(a^j,\xi^j,w) = h(a,\xi_0,w)$$

$$= \frac{i\ell^*}{2}[h_n(w) + \overline{h_n(w)}] - (n-1)\sum_{\beta,\gamma=1}^{n} a_\beta(\psi_{\beta\gamma}h_\gamma(w) + \psi_{\beta\bar\gamma}\overline{h_\gamma(w)})$$

$$+ \frac{i(n-1)}{2}a_n\sum_{\beta,\gamma=1}^{n}\{\psi_{\beta\gamma}h_{\beta\gamma}(w) + \psi_{\bar\beta\bar\gamma}\overline{h_{\beta\gamma}(w)} + \psi_{\beta\bar\gamma}h_{\beta\bar\gamma}(w) + \psi_{\bar\beta\gamma}h_{\bar\beta\gamma}(w)\}$$

$$+ \frac{(n-1)}{2}[h_n(w) - \overline{h_n(w)}]\left[\sum_{\beta=1}^{n} a_\beta(\psi_{\beta n} - \psi_{\beta\bar n})\right.$$

$$\left. + \frac{a_n}{i(n-1)}\frac{i}{2(2n-1)}(\Delta\psi + (n-1)\psi_{n\bar n})\right].$$

Clearly, then, $h(a,\xi_0,w)$ belongs to the linear span of $\mathcal{E}$ so that

(12.9) $$h(a,\xi_0,w) \equiv 0 \quad \text{for} \quad w \in D(\xi_0)$$

holds. We now analyze (12.11) and its behavior with $a = (a_1,\ldots,a_n)$.

Case 1. $a_n \ne 0$.

By hypothesis, either (12.5) or (12.6) holds. Hence there exists α,β with $1 \le \alpha,\beta \le n-1$ such that either $\psi_{\alpha\beta} \equiv \frac{\partial^2\psi}{\partial\xi_\alpha\partial\xi_\beta}(\xi_0) \ne 0$ or $\psi_{\alpha\bar\beta} \equiv \frac{\partial^2\psi}{\partial\xi_\alpha\partial\bar\xi_\beta}(\xi_0) \ne 0$. From (12.11), this implies that either the coefficient of $h_{\alpha\beta}(w)$ or the coefficient of $h_{\alpha\bar\beta}(w)$ is non-zero. Hence, from the linear independence of the set $\mathcal{E}$, $h(a,\xi_0,w) \ne 0$ for $w \in D(\xi_0)$, contradicting (12.9).

<u>Case</u> <u>2</u>. $a_n = 0$.

Using (12.11), we obtain

(12.12)

$$h(a,\xi_0,w) = \left[\frac{i\ell^*}{2} - \left(\frac{n-1}{2} \right) \sum_{\beta=1}^{n-1} a_\beta(\psi_{\beta n} + \psi_{\beta\bar{n}}) \right][h_n(w) + \overline{h_n(w)}]$$

$$- (n-1) \sum_{\gamma=1}^{n-1} \left[\left(\sum_{\beta=1}^{n-1} a_\beta \psi_{\beta\gamma} \right) h_\gamma(w) + \left(\sum_{\beta=1}^{n-1} a_\beta \psi_{\beta\bar{\gamma}} \right) \overline{h_\gamma(w)} \right] .$$

Since either $[\psi_{\beta\gamma}]_{\beta,\gamma=1,\ldots,n-1}$ or $[\psi_{\beta\bar{\gamma}}]_{\beta,\gamma=1,\ldots,n-1}$ is nonsingular and $(a_1,\ldots,a_{n-1}) \neq 0$ either $\sum_{\beta=1}^{n-1} a_\beta\psi_{\beta\gamma} \neq 0$ or $\sum_{\beta=1}^{n-1} a_\beta\psi_{\beta\bar{\gamma}} \neq 0$ for some $\gamma \in \{1,\ldots,n-1\}$. Hence either the coefficient of $h_\gamma(w)$ or the coefficient of $\overline{h_\gamma(w)}$ is nonzero. Again, from the linear independence of $\mathcal{E}$, $h(a,\xi_0,w) \not\equiv 0$ for $w \in D(\xi_0)$, contradicting (12.9).

Hence $m > 0$ and 1 is proved.

2., 3. Recalling the definitions, for $\rho > 0$, of

$$M_\rho \equiv \{(\xi,a) \in D \times S_n\colon \ |\ell_1(\xi,a)| < \rho\}$$

and

$$M'_\rho \equiv \{(\xi,a) \in D \times S_n\colon \ |\ell_1(\xi,a)| \geq \rho\}$$

it follows from 1. that

$$\ell_2(\xi,a) \geq \frac{m}{\rho^2} |\ell_1(\xi,a)|^2$$

for $(\xi,a) \in M_\rho$ and any $\rho > 0$. If we choose ρ_0, $c > 0$ as in Corollary 9.2 so that

$$\ell_2(\xi,a) \geq \left(\frac{c}{2} \right)^2 |\ell_1(\xi,a)|^2 \tag{9.14}$$

for $(\xi, a) \in M'_{\rho_0}$, then setting

$$\tilde{c} \equiv \min \left\{ \left(\frac{c}{2} \right)^2 , \frac{m}{\rho_0^2} \right\}$$

implies that

$$\ell_2(\xi, a) \geq \tilde{c} \, |\ell_1(\xi, a)|^2$$

for all $(\xi, a) \in D \times S_n$. Hence 2. is proved and 3. follows from Lemma 9.1. ∎

Remark. Under the hypothesis of Theorem 12.1, it follows that (D, ds^2) satisfies the separation property as defined in Chapter 10 where ds^2 is the Λ-metric for D .

13. AN EXAMPLE WITH $\ell_2(\xi,a) \not\geq c|\ell_1(\xi,a)|^2$

In our final chapter, we give an example of a pseudoconvex domain $D \subset \mathbb{C}^2$ with smooth boundary for which there does not exist a constant $c > 0$ such that

(13.1) $$\ell_2(\xi,a) \geq c|\ell_1(\xi,a)|^2$$

for all $(\xi,a) \in D \times S_n$. We begin by defining

$$E = \{z = (z_1,z_2) \in \mathbb{C}^2: \ \psi(z_1,z_2) < 0\}$$

where

$$\psi(z_1,z_2) = 2\,\mathrm{Re}\left(iz_2 + z_1z_2 - \frac{z_1^3}{3} \right)$$
$$= iz_2 - i\bar{z}_2 + z_1z_2 + \overline{z_1z_2} - \frac{z_1^3}{3} - \frac{\bar{z}_1^3}{3}$$
$$= -2x_4 + 2x_1x_3 - 2x_2x_4 - \frac{2}{3}x_1^3 + 2x_1x_2^2$$

and $z_1 = x_1 + ix_2$, $z_2 = x_3 + ix_4$. Note that $(0,0) \in \partial E$ and

(13.2)
$$\mathrm{Grad}_{(z)}\psi(z_1,z_2) = (z_2 - z_1^2, i + z_1); \text{ in particular, } \mathrm{Grad}_{(z)}\psi(0,0) = (0,i).$$

The complex tangent space to ∂E at $(0,0)$ is thus given by

$$\pi = \{(z_1,z_2) \in \mathbb{C}^2: \ z_2 = 0\}$$

and we have the complex Hessian of ψ is

(13.3) $$\begin{pmatrix} \frac{\partial^2\psi}{\partial z_1\partial\bar{z}_1}(z_1,z_2) & \frac{\partial^2\psi}{\partial z_1\partial\bar{z}_2}(z_1,z_2) \\ \frac{\partial^2\psi}{\partial z_2\partial\bar{z}_1}(z_1,z_2) & \frac{\partial^2\psi}{\partial z_2\partial\bar{z}_2}(z_1,z_2) \end{pmatrix} \equiv \begin{pmatrix} 0 & 0 \\ 0 & 0 \end{pmatrix}$$

Hence E is Levi-flat. An elementary computation of the real Hessian of ψ shows that ∂E is not convex at $(0,0)$. Furthermore, since

$$(13.4)\qquad \begin{pmatrix} \frac{\partial^2\psi}{\partial z_1 \partial z_1}(z_1,z_2) & \frac{\partial^2\psi}{\partial z_1 \partial z_2}(z_1,z_2) \\ \frac{\partial^2\psi}{\partial z_2 \partial z_1}(z_1,z_2) & \frac{\partial^2\psi}{\partial z_2 \partial z_2}(z_1,z_2) \end{pmatrix} = \begin{pmatrix} -2z_1 & 1 \\ 1 & 0 \end{pmatrix},$$

$\frac{\partial^2\psi}{\partial z_1 \partial z_1}(0,0) = 0$ so that $(0,0)$ is not a nondegenerate boundary point of E. To obtain a bounded pseudoconvex domain with the same properties as E at $(0,0)$, we let

$$B = \{(z_1,z_2): \ |z_1|^2 + |z_2|^2 < 1\}$$

be the unit ball in $\mathbb{C}^2$ and consider $D \equiv E \cap B$. Then D is a pseudoconvex domain with corners $e \equiv \partial E \cap \partial B$; by modifying D near e we obtain a smoothly bounded pseudoconvex domain, which we again call D, for which $(0,0)$ is neither a convex, strictly pseudoconvex, nor nondegenerate boundary point. In particular, we cannot appeal to either Theorem 9.1 (convex case) or Theorem 12.1 (2.) (strictly pseudoconvex or non-degenerate case) to conclude that (13.1) holds for some $c > 0$; indeed, we show that with the choice of $a = (1,0) \in S_2$, inequality (13.1) <u>fails</u> to hold in D for any $c > 0$ by analyzing the behavior of $\ell_2(\xi,a)$ and $\ell_1(\xi,a)$ for ξ near $(0,0)$.

To achieve our goal, we begin by defining, for $\varepsilon > 0$, $\eta_\varepsilon \equiv (\sqrt{i\varepsilon}, i\varepsilon) \in \mathbb{C}^2$ (we take $\sqrt{k} > 0$ if $k > 0$). For ε sufficiently small,

$$\psi(\eta_\varepsilon) = -2\varepsilon - \frac{2\sqrt{2}}{3}\varepsilon\sqrt{\varepsilon} < 0$$

so that $\eta_\varepsilon \in D$. Clearly

$$\lim_{\varepsilon\to 0} \eta_\varepsilon = (0,0) \in \partial D .$$

If we set $a = (1,0)$, then a lies in the complex tangent space to ∂D at $(0,0)$ by (13.2). Thus Proposition 11.1 does not apply to bound the ratio

$$\ell_2(\eta_\varepsilon, a)/|\ell_1(\eta_\varepsilon, a)|^2$$

away from 0 for small ε. We show that

$$\lim_{\varepsilon\to 0} \ell_1(\eta_\varepsilon, a) = -i \tag{13.5}$$

and

$$\lim_{\varepsilon\to 0} \ell_2(\eta_\varepsilon, a) = 0 \tag{13.6}$$

and hence we cannot find $c > 0$ so that (13.1) holds for $(\xi, a) \in D \times S_2$.

To prove (13.5), since $a = (1,0)$ and $\lambda(\xi) = \Lambda(\xi)\psi(\xi)^2$,

$$\ell_1(\xi, a) = \frac{\frac{\partial\Lambda}{\partial\xi_1}(\xi)}{\Lambda(\xi)} = \frac{\frac{\partial\lambda}{\partial\xi_1}(\xi)}{\lambda(\xi)} - 2\,\frac{\frac{\partial\psi}{\partial\xi_1}(\xi)}{\psi(\xi)} \quad \text{where} \quad \xi = (\xi_1, \xi_2) .$$

From (13.2), $\frac{\partial\psi}{\partial\xi_1}(\xi) = \xi_2 - \xi_1^2$ so that

$$\lim_{\varepsilon\to 0} \frac{\frac{\partial\psi}{\partial\xi_1}(\eta_\varepsilon)}{\psi(\eta_\varepsilon)} = \lim_{\varepsilon\to 0} \frac{i\varepsilon - i\varepsilon}{-2\varepsilon - \frac{2\sqrt{2}}{3}\,\varepsilon\sqrt{\varepsilon}} = 0$$

and

$$\lim_{\varepsilon\to 0} \ell_1(\eta_\varepsilon, a) = \lim_{\varepsilon\to 0} \frac{\frac{\partial\lambda}{\partial\xi_1}(\eta_\varepsilon)}{\lambda(\eta_\varepsilon)} = \frac{\frac{\partial\lambda}{\partial\xi_1}(0,0)}{\lambda(0,0)} = -\,i(\psi_{12} - \psi_{1\bar{2}})$$

from (11.14) and $\lambda(0,0) = -\|\text{Grad}\psi(0,0)\|^2 = -1$. Since

$$\psi_{12} \equiv \frac{\partial^2\psi}{\partial z_1 \partial z_2}(0,0) = 1 \quad \text{and} \quad \psi_{1\bar{2}} \equiv \frac{\partial^2\psi}{\partial z_1 \partial \bar{z}_2}(0,0) = 0$$

from (13.3) and (13.4), we obtain (13.5).

To prove (13.6), note first that by using (12.12) with $\ell^* = -i$,

$$h(a,(0,0),w) = \left[\frac{1}{2} - \frac{1}{2}(\psi_{12} + \psi_{1\bar{2}})\right]\left[h_2(w) + \overline{h_2(w)}\right] - \psi_{11}h_1(w) - \psi_{1\bar{1}}\overline{h_1(w)}$$

$$\equiv 0 \quad \text{for} \quad w \in D(0)$$

since $\psi_{12} = 1$ and $\psi_{11} = \psi_{1\bar{1}} = \psi_{1\bar{2}} = 0$ from (13.3) and (13.4). Hence

$$\tilde{J}((0,0),a) \equiv \frac{4}{w_4} \iint_{D(0)} \sum_{\alpha=1}^{2} \left|\frac{\partial}{\partial \bar{w}_\alpha} h(a,(0,0),w)\right|^2 dV_w = 0 .$$

Since $\lim_{\varepsilon\to 0} \eta_\varepsilon = (0,0) \in \partial D$, by using arguments similar to those used in Chapter 5, we can show that $\lim_{\varepsilon\to 0} \tilde{J}(\eta_\varepsilon, a) = \tilde{J}((0,0),a)$. Hence

(13.7) $$\lim_{\varepsilon\to 0} \tilde{J}(\eta_\varepsilon, a) = 0 .$$

To finish the proof of (13.6), we must show that

(13.8) $$\lim_{\varepsilon\to 0} \tilde{I}(\eta_\varepsilon, a) = 0$$

where

(13.9) $$\tilde{I}(\eta_\varepsilon, a) = \frac{1}{w_4(-\Lambda(\eta_\varepsilon))} \int_{\partial D} K_2(z,\theta) \|\mathrm{Grad}_{(\bar{z})} G(\eta_\varepsilon, z)\| dS_z$$

and for $z \in \partial D$,

(13.10) $$\theta = \theta(a,\eta_\varepsilon,z) = -\ell_1(\eta_\varepsilon, a) \cdot (z - \eta_\varepsilon) + a ;$$

while

$$(13.11)\qquad K_2(z,\theta) \equiv \frac{1}{\|\mathrm{Grad}_{(z)}\tilde{\psi}\|^3}\left[\sum_{\alpha,\beta=1}^{2} \frac{\partial^2\tilde{\psi}}{\partial z_\alpha \partial \bar{z}_\beta}\theta_\alpha\bar{\theta}_\beta \|\mathrm{Grad}_{(z)}\tilde{\psi}\|^2 \right.$$

$$- 2\,\mathrm{Re}\left\{\sum_{i,\alpha,\beta=1}^{2} \frac{\partial\tilde{\psi}}{\partial z_\alpha}\frac{\partial^2\tilde{\psi}}{\partial \bar{z}_\beta \partial z_i}\frac{\partial\tilde{\psi}}{\partial z_i}\theta_\alpha\bar{\theta}_\beta\right\}$$

$$\left. + \left|\sum_{\alpha=1}^{2}\frac{\partial\tilde{\psi}}{\partial z_\alpha}\theta_\alpha\right|^2 \Delta_{(z)}\tilde{\psi}\right] .$$

where $\tilde{\psi}$ is a smooth defining function for D . Using $a = (1,0)$ and $\xi = \eta_\varepsilon$ in (13.10) , equation (13.5) implies that

$$\lim_{\varepsilon\to 0} \theta(a,\eta_\varepsilon,z) = i(z_1,z_2) + (1,0) = (iz_1 + 1, iz_2) \quad \text{for} \quad z = (z_1,z_2) \in \partial D .$$

Since D is bounded this implies that there exists M such that

$$(13.12)\qquad \|\theta(a,\eta_\varepsilon,z)\| < M$$

for $z \in \partial D$ if ε is sufficiently small.

We write ∂D as the disjoint union of the three sets S_1, S_2, S_3 where

S_1 = modified portion of ∂D near $e \equiv \partial E \cap \partial B$;

$S_2 = \partial E \cap B - S_1$;

$S_3 = \partial B \cap E - S_1$.

By (13.12), we can find M_1 such that

$$(13.13)\qquad \int_{S_1} K_2(z,\theta)\|\mathrm{Grad}_{(z)}G(\eta_\varepsilon,z)\|dS_z < M_1$$

for ε sufficiently small. On S_2 , we can take

$$\tilde{\psi}(z_1,z_2) = \psi(z_1,z_2) = 2\,\mathrm{Re}\left(iz_2 + z_1z_2 - \frac{z_1^3}{3}\right),$$

i.e., the defining function for E . Hence $K_2(z,\theta) \equiv 0$ for $z \in S_2$.

On S_3, we can take

$$\tilde{\psi}(z_1, z_2) = |z_1|^2 + |z_2|^2 - 1 ,$$

i.e., the defining function for B. Direct computation shows that

$$K_2(z,\theta) \equiv 4\left[|\theta_1|^2 + |\theta_2|^2 - 2 \operatorname{Re} \{z_1\theta_1\bar{z}_2\bar{\theta}_2\} \right]$$

for $z \in S_3$, hence, by (13.12), we can find M_2 such that

(13.14) $$K_2(z,\theta) < M_2$$

for $z \in S_3$, $\theta = \theta(a, \eta_\varepsilon, z)$, if ε is sufficiently small. Using (13.13) and (13.14) in (13.9), we obtain

$$\tilde{I}(\eta_\varepsilon, a) \le \frac{1}{w_4(-\Lambda(\eta_\varepsilon))}\left[M_1 + M_2 \int_{S_3} \|\mathrm{Grad}_{(z)} G(\eta_\varepsilon, z)\|^2 dS_z \right]$$

Since $\lim_{\varepsilon\to 0} [-\Lambda(\eta_\varepsilon)] = -\Lambda(0,0) = +\infty$, to prove (13.8), it suffices to show that

(13.15) $$\lim_{\varepsilon\to 0} \frac{1}{-\Lambda(\eta_\varepsilon)} \int_{S_3} \|\mathrm{Grad}_{(z)} G(\eta_\varepsilon, z)\|^2 dS_z = 0$$

To prove (13.15), let

$$G_0(\eta_\varepsilon, z) = \frac{1}{\|z-\eta_\varepsilon\|^2} - \frac{1}{\|\eta_\varepsilon\|^2} \frac{1}{\left\| z - \dfrac{\eta_\varepsilon}{\|\eta_\varepsilon\|^2} \right\|^2}$$

be the Green function for (B, η_ε). By the maximum principle,

$$G(\eta_\varepsilon, z) < G_0(\eta_\varepsilon, z) \quad \text{for} \quad z \in D .$$

Since $G(\eta_\varepsilon, z) = G_0(\eta_\varepsilon, z) = 0$ for $z \in S_3$,

$$\|\mathrm{Grad}_{(z)} G(\eta_\varepsilon, z)\| \le \|\mathrm{Grad}_{(z)} G_0(\eta_\varepsilon, z)\| \quad \text{for} \quad z \in S_3 .$$

If η_ε is sufficiently close to 0, it follows by direct computation that

$$\|\mathrm{Grad}_{(z)} G_0(\eta_\varepsilon, z)\| \le 2 \quad \text{for} \quad z \in S_3 .$$

Since $\lim_{\varepsilon\to 0} \Lambda(\eta_\varepsilon) = -\infty$, (13.15) follows.

Remark. We do not know if the Λ-metric is complete for the domain D considered in this chapter. Indeed, it remains an open problem to determine whether the Λ-metric ds^2 is complete for an arbitrary bounded pseudoconvex domain $D \subset \mathbb{C}^n$ with smooth boundary. We mention that if D is a bounded pseudoconvex domain with C^1 boundary, Ohsawa [OH] has shown that the Bergman metric $ds_B^2 \equiv \sum_{\alpha,\beta=1}^{n} \frac{\partial^2 \log K}{\partial \xi_\alpha \partial \bar{\xi}_\beta} d\xi_\alpha \otimes d\bar{\xi}_\beta$ (here, $K = K(\xi, \xi)$ = Bergman kernel for D on the diagonal $z = \xi$; cf. chapter 8) is complete for D. This metric is invariant under biholomorphic mappings of D; it is unclear how the Λ-metric is related to biholomorphic invariants of D.

REFERENCES

[BT] E. Bedford and B.A. Taylor, Plurisubharmonic functions with logarithmic singularities, Annales Inst. Fourier, Grenoble, 38, 4 (1988), 133-171.

[F] C.L. Fefferman, The Bergman kernel and biholomorphic mappings of pseudoconvex domains, Invent. Math., 26 (1974), 1-65.

[H] L.L. Helms, Introduction to Potential Theory. Wiley-Interscience (1969), New York.

[K] S. Krantz, Function Theory of Several Complex Variables, Wiley-Interscience (1982), New York.

[N] T. Nishino, Nouvelles recherches sur les fonctions entieres de plusieurs variables complexes (I)-(V), J. Math. Kyoto Univ., 8 (1968), 49-100; 9 (1969), 221-224; 10 (1970), 245-271; 13 (1973), 217-272; 15 (1975), 527-553.

[OH] T. Ohsawa, A remark on the completeness of the Bergman metric, Proc. Japan Acad. Ser. A Math. Sci., 57 (1981), no. 4, 238-240.

[OK] K. Oka, Note sur les familles de fonctions analytiques multiformes, etc., J. Sci. Hiroshima Univ., A-4 (1934), 93-98.

[O] R. Osserman, A Survey of Minimal Surfaces, Van-Nostrand (1969), New York.

[PW] M.H. Protter and H.F. Weinberger, Maximum Principles in Differential Equations, Prentice-Hall (1967), New Jersey.

[R] H.E. Rauch, Weierstrass points, branch points, and the moduli of Riemann surfaces, Comm. Pure and Applied Math., 12 (1959), 543-560.

[S] N. Suita, Capacities and kernels on Riemann surfaces, Arch. Rational Mech. Anal., 46 (1972), 212-217.

[Y_1] H. Yamaguchi, Parabolicité d'une fonction entière, J. Math. Kyoto Univ., 16 (1976), 71-92.

[Y] H. Yamaguchi, Variations of pseudoconvex domains over $\mathbb{C}^n$, Mich. Math. Journal, 36 (1989), 415-457.

Department of Mathematics
Wellesley College
Wellesley, MA 02181
USA

Department of Mathematics
Shiga University
Otsu-City, Shiga 520
JAPAN

AMS/MOS Classification numbers
Primary: 32F05
Secondary: 31C10, 32F15

MEMOIRS of the American Mathematical Society

SUBMISSION. This journal is designed particularly for long research papers (and groups of cognate papers) in pure and applied mathematics. The papers, in general, are longer than those in the TRANSACTIONS of the American Mathematical Society, with which it shares an editorial committee. Mathematical papers intended for publication in the Memoirs should be addressed to one of the editors:

Ordinary differential equations, partial differential equations and applied mathematics to ROGER D. NUSSBAUM, Department of Mathematics, Rutgers University, New Brunswick, NJ 08903

Harmonic analysis, representation theory and Lie theory to AVNER D. ASH, Department of Mathematics, The Ohio State University, 231 West 18th Avenue, Columbus, OH 43210

Abstract analysis to MASAMICHI TAKESAKI, Department of Mathematics, University of California, Los Angeles, CA 90024

Real and harmonic analysis to DAVID JERISON, Department of Mathematics, M.I.T., Rm 2-180, Cambridge, MA 02139

Algebra and algebraic geometry to JUDITH D. SALLY, Department of Mathematics, Northwestern University, Evanston, IL 60208

Geometric topology and general topology to JAMES W. CANNON, Department of Mathematics, Brigham Young University, Provo, UT 84602

Algebraic topology and differential topology to RALPH COHEN, Department of Mathematics, Stanford University, Stanford, CA 94305

Global analysis and differential geometry to JERRY L. KAZDAN, Department of Mathematics, University of Pennsylvania, E1, Philadelphia, PA 19104-6395

Probability and statistics to RICHARD DURRETT, Department of Mathematics, Cornell University, Ithaca, NY 14853-7901

Combinatorics and number theory to CARL POMERANCE, Department of Mathematics, University of Georgia, Athens, GA 30602

Logic, set theory, general topology and universal algebra to JAMES E. BAUMGARTNER, Department of Mathematics, Dartmouth College, Hanover, NH 03755

Algebraic number theory, analytic number theory and modular forms to AUDREY TERRAS, Department of Mathematics, University of California at San Diego, La Jolla, CA 92093

Complex analysis and nonlinear partial differential equations to SUN-YUNG A. CHANG, Department of Mathematics, University of California at Los Angeles, Los Angeles, CA 90024

All other communications to the editors should be addressed to the Managing Editor, DAVID J. SALTMAN, Department of Mathematics, University of Texas at Austin, Austin, TX 78713.

General instructions to authors for

PREPARING REPRODUCTION COPY FOR MEMOIRS

For more detailed instructions send for AMS booklet, "A Guide for Authors of Memoirs." Write to Editorial Offices, American Mathematical Society, P.O. Box 6248, Providence, R.I. 02940.

MEMOIRS are printed by photo-offset from camera copy fully prepared by the author. This means that the finished book will look exactly like the copy submitted. Thus the author will want to use a good quality typewriter with a new, medium-inked black ribbon, and submit clean copy on the appropriate model paper.

Model Paper, provided at no cost by the AMS, is paper marked with blue lines that confine the copy to the appropriate size.

Special Characters may be filled in carefully freehand, using dense black ink, or **INSTANT** ("rub-on") **LETTERING** may be used. These may be available at a local art supply store.

Diagrams may be drawn in black ink either directly on the model sheet, or on a separate sheet and pasted with rubber cement into spaces left for them in the text. Ballpoint pen is not acceptable.

Page Headings (Running Heads) should be centered, in CAPITAL LETTERS (preferably), at the top of the page — just above the blue line and touching it.

LEFT-hand, EVEN-numbered pages should be headed with the AUTHOR'S NAME;

RIGHT-hand, ODD-numbered pages should be headed with the TITLE of the paper (in shortened form if necessary).

Exceptions: PAGE 1 and any other page that carries a display title require NO RUNNING HEADS.

Page Numbers should be at the top of the page, on the same line with the running heads.

LEFT-hand, EVEN numbers — flush with left margin;

RIGHT-hand, ODD numbers — flush with right margin.

Exceptions: PAGE 1 and any other page that carries a display title should have page number, centered below the text, on blue line provided.

FRONT MATTER PAGES should be numbered with Roman numerals (lower case), positioned below text in same manner as described above.

MEMOIRS FORMAT

It is suggested that the material be arranged in pages as indicated below.
Note: <u>Starred items</u> (*) <u>are requirements of publication.</u>

Front Matter (first pages in book, preceding main body of text).

Page i — *Title, *Author's name.

Page iii — Table of contents.

Page iv — *Abstract (at least 1 sentence and at most 300 words).

Key words and phrases, if desired. (A list which covers the content of the paper adequately enough to be useful for an information retrieval system.)

*<u>*1991 Mathematics Subject Classification*</u>. This classification represents the primary and secondary subjects of the paper, and the scheme can be found in Annual Subject Indexes of MATHEMATICAL REVIEWS beginnning in 1990.

Page 1 — Preface, introduction, or any other matter not belonging in body of text.

Footnotes: *Received by the editor date.
Support information — grants, credits, etc.

First Page Following Introduction - Chapter Title (dropped 1 inch from top line, and centered). Beginning of Text.

Last Page (at bottom) - Author's affiliation.